INTERNATIONAL SERIES OF MONOGRAPHS ON
NUCLEAR ENERGY
GENERAL EDITOR: J. V. DUNWORTH

Division II: NUCLEAR PHYSICS

Volume 5

THE RELATION BETWEEN THE PHOTO-PRODUCTION AND THE SCATTERING OF π-MESONS

OTHER TITLES IN THE INTERNATIONAL SERIES OF MONOGRAPHS ON NUCLEAR ENERGY

DIVISION I. ECONOMICS AND LAW

Vol. 1—Atomic Power: An Appraisal—C. ALLARDICE (Ed.)

DIVISION II. NUCLEAR PHYSICS

Vol. 1—Neutron Cross Sections—D. J. HUGHES.
Vol. 2—Physics of Nuclear Fission—J. E. S. BRADLEY (Trans.).
Vol. 3—Soviet Reviews of Nuclear Science—Translation of "Atomnaya Energiya", vol. 3, no. 11 (1957).
Vol. 4—Fast Reactor Cross Sections—S. YIFTAH, D. OKRENT and P. A. MOLDAUER.

DIVISION IV. ISOTOPES AND RADIATION

Vol. 1—Atlas of Gamma-ray Spectra from Radiative Capture of Thermal Neutrons—L. V. GROSHEV, V. N. LUTSENKO, A. M. DEMIDOV and V. I. PELEKHOV.

DIVISION V. HEALTH PHYSICS

Vol. 1—Health Physics Instrumentation—J. S. HANDLOSER.
Vol. 2—Radioactive Wastes: Treatment and Disposal—C. B. AMPHLETT.

DIVISION VI. MEDICINE

Vol. 1—Radioisotope Studies of Fatty Acid Metabolism—J. F. MEAD and D. R. HOWTON.

DIVISION VII. REACTOR ENGINEERING

Vol. 1—Shielding Materials for Nuclear Reactors—A. N. KOMAROVSKII.

DIVISION X. REACTOR DESIGN PHYSICS

Vol. 1—An Introduction to Reactor Physics—D. J. LITTLER and J. F. RAFFLE.
Vol. 2—Radiation Shielding—B. T. PRICE, C. C. HORTON and K. T. SPINNEY.
Vol. 3—Thermal Reactor Theory—A. D. GALANIN.
Vol. 4—Resonance Absorption in Nuclear Reactors—L. DRESNER.

DIVISION XIV. PLASMA PHYSICS AND THERMONUCLEAR RESEARCH

Vol. 1—An Introduction to Thermonuclear Research—A. SIMON.

The Relation Between the Photo-production and the Scattering of π-mesons

by

N. F. NELIPA

Translated by

PRASENJIT BASU

PERGAMON PRESS
NEW YORK · OXFORD · LONDON · PARIS
1961

PERGAMON PRESS INC.
122 East 55th Street, New York 22, N.Y.
1404 New York Avenue N.W., Washington 5 D.C.
Statler Center 640, 900 Wilshire Boulevard
Los Angeles 17, California

PERGAMON PRESS LTD.
Headington Hill Hall, Oxford
4 & 5 Fitzroy Square, London W.1

PERGAMON PRESS S.A.R.L.
24 Rue des Écoles, Paris Ve

PERGAMON PRESS G.m.b.H.
Kaiserstrasse 75, Frankfurt am Main

A translation of the original volume
Svyaz' fotoobrazovaniya π-mezonov s rasseyaniyem
(Moscow, Atomizdat, 1959)

Library of Congress Card Number: 61–15280

Printed in Great Britain by Pergamon Printing and Art Services Limited, London

CONTENTS

PREFATORY NOTE

This book provides a systematic review of researches on the relationship between the photoproduction and the scattering of π-mesons and on the application of the results obtained to the analysis of experimental data.

The book is intended for scientific workers studying nuclear reactions at high energies, and for students attending advanced courses of nuclear physics.

INTRODUCTION

Extensive experimental material has lately been obtained on phenomena in which mesons[1] take part, in particular on processes of scattering and photo-production of mesons on nucleons. There is, however, no consistent detailed theory as yet, with the help of which this experimental material could be analysed. For this reason, methods based on a few general hypotheses (laws of conservation, isotopic invariance, unitary nature of the S-matrix, its reversibility in time etc.) and some quite natural physical assumptions (short-range nuclear forces etc.) acquire special interest.

Experimental data on the scattering of mesons with low energies ($\leqslant$ 300 Mev) on nucleons can be analysed if it is assumed that

1) the nuclear forces are short-range forces

2) the hypothesis of charge independence is correct

3) the predominant role in scattering is played by the state of the meson-nucleon system with total isotopic spin $I = 3/2$ and total angular momentum $J = 3/2$ (i.e. in this state the interaction between the meson and the nucleon is particularly strong).

But the analysis of experimental data on the photo-production of mesons on nucleons within the framework of the above model is complicated by a certain fact. The number of unknown quantities entering the differential cross-section is so great that the experimental data on photo-production alone is found to be insufficient for their determination. Here it is possible to use some relationships connecting the characteristics of the processes of scattering and photo-production. It can be shown that the known

1 Henceforth the term "meson" will always be used to mean π-meson.

phases of scattering of mesons on nucleons will be the arguments of the unknown complex matrix elements of photo-production entering the coefficients of the angular variables in the differential cross-section. Then only the moduli of the matrix elements will remain as unknowns in the expression for the cross-section of photo-production. It is already possible to determine some of these with the help of the available experimental data on the photo-production of mesons.

A large number of mathematical and physical papers have been devoted to this subject in the past few years, but these papers are scattered in various journals and are often written in a very compressed form. We therefore thought it worthwhile to describe systematically and with as much detail as possible the results contained in these papers.

The method described is applicable to γ-quanta of any energy, but it becomes complicated with increase in energy because of the increase in the number of partial waves which have to be taken into consideration. If we confine ourselves to the case where mesons are formed mainly in the s- and p- states, the analysis can be carried out, roughly speaking, up to $E_\gamma \leqslant 400$ Mev. We shall confine ourselves to a consideration of this range of energy. First of all, we shall describe the experimental data and derive the expression for the differential cross-section of the photo-production of mesons. We then proceed to establish a relation between the processes mentioned and discuss the application of the results obtained to the analysis of experimental data on the photo-production of mesons.

We have not aimed at comprehensiveness in the exposition. Rather, we have deliberately chosen the most thoroughly studied processes of scattering and photo-production of π-mesons on nucleons and tried to illustrate with as much as detail as possible, with the help of this concrete example, how the method works. We feel that after the reader has mastered the material in this book he will not find it difficult to follow a similar analysis of other nuclear processes.

The calculations are simplified to a great extent by tables of numerical values of Clebsch-Gordan, Racah and Z-coefficients. We have therefore given tables of the most commonly used values of these coefficients, which can be used for the study of other processes also.

1. EXPERIMENTAL DATA

Scattering of mesons. The following processes of scattering of mesons on protons have been studied (1, 2, 3);

$$\pi^+ + p \rightarrow \pi^+ + p; \tag{1a}$$

$$\pi^- + p \rightarrow \pi^0 + n; \tag{1b}$$

$$\pi^- + p \rightarrow \pi^- + p. \tag{1c}$$

In the energy interval in which we are interested ($\leqslant$300 Mev) we can confine ourselves to a consideration of scattering only in the *s*- and *p*-states (1a). Then the angular distribution of each of these processes (1b) in the centre-of-mass system can be written in the form

$$\sigma(\vartheta) = a + b\cos\vartheta + c\cos^2\vartheta. \tag{1.2}$$

Each of the processes (1) is characterized by its own values of the coefficients *a*, *b*, *c*, determined by the energy of the meson.

The experimental values of the coefficients *a*, *b*, *c* for each of the processes (1) for a given energy are determined from the corresponding experimental curves characterizing the angular distribution of the scattered mesons. On the other hand, theoretical expressions containing the phases of scattering can be found for the same coefficients *a*, *b*, *c* (1, 4). A comparison of the respective experimental and theoretical values of *a*, *b*, *c* makes it possible, in principle, to determine the phases of scattering. If the hypothesis of charge independence is used, then for the case where *s*- and *p*-waves are considered the processes (1) can be described with the help of six phases α_1, α_3, α_{33}, α_{31}, α_{13}, α_{11} (the first suffix corresponds to twice the value of the isotopic spin, and the second to twice the orbital angular momentum of the meson-nucleon system[1]).

Calculation of the scattering phases by this method leads to a number of ambiguities. Firstly, it is found that if nuclear forces alone are considered the signs of the scattering phases cannot be determined. In this case only the squares of the sines of the scattering phases enter the expressions for *a, b, c* (1, 4). This ambiguity may be removed (5, 6) with the help of experiments on the scattering of slow mesons ($\leqslant$100 Mev) at small angles, when the Coulomb interaction also becomes considerable. The interference of the Coulomb interaction with the nuclear interaction leads to the appearance of a term containing the meson scattering phases in the first degree. With the help of the experiments mentioned, it has been established that $\alpha_{33} > 0$, $\alpha_1 > 0$ and $\alpha_3 < 0$.

Secondly, in the phase analysis two different sets of values of the scattering phases were obtained which satisfy the experimental data equally well. These sets differ from one another in that in one of them ("Fermi's solution") the maximum contribution to scattering comes from the phase α_{33}, while in the other ("Yang's solution") it comes from the phase α_{31}. There are a number of arguments to support the belief that the correct solution is that of Fermi and not that of Yang. This follows from an analysis of the experimental data on scattering with the help of the dispersion relations (7), and also from the relationships connecting the processes of scattering and photo-production (1) etc.

Finally, due to the large experimental errors in the phase analysis of experimental data on the scattering of mesons, an additional set of phases is obtained for each energy. But the majority of these sets can be rejected because they do not satisfy a number of essential requirements (do not give the correct value for the total cross-section, sharply change with energy variation etc.).

Thus on the basis of the considerations put forward we can give preference to that set of phases in which the principal part is played by the positive phase α_{33}, and the phases α_1 and α_3 have different values for small energies.

---

1 Henceforth simply α_1 and α_3 will be written instead of α_{10} and α_{30}.

The values of the scattering phases for various energies are given in Table 1 (3). The phases α_{11}, α_{13} and α_{31} may be left out of consideration since they are considerably less than the phases α_{33}, α_3 and α_1.

TABLE 1

Phases of scattering (in degrees)

E_γ	α_{33}	α_1	α_3
200	5.9	7.5	— 4.0
230	13.5	9.5	— 7.5
260	24.8	9.2	—10.7
290	42.0	6.6	—14.0
320	66.5	3.0	—16.5
350	92.0	—1.5	—19.0
380	117.0	—6.0	—21.5
410	142.0	—10.0	—24.0

REMARK. On the subject of more recent data on the phase analysis of the scattering of π-mesons see the review article presented by B.M. Pontecorvo at the Kiev conference on the Physics of High Energy Particles, 1959.

Photo-production of mesons on nucleons. Photo-production of mesons on nucleons may occur in the following four processes:

$$\gamma + p \rightarrow n + \pi^+; \tag{2a}$$

$$\gamma + p \rightarrow p + \pi^0; \tag{2b}$$

$$\gamma + n \rightarrow p + \pi^-; \tag{2c}$$

$$\gamma + n \rightarrow n + \pi^0. \tag{2d}$$

Let us assume (3), that γ-quanta with an energy $E_\gamma \leqslant 400$ Mev form mesons only in the *s*- and *p*-states[1]. Then the angular distribution of the mesons in the CM system can be denoted, as in

1 See below on the role of the *d*-phase in this energy region.

the case of scattering, by the expression

$$\sigma(\vartheta) = A + B\cos\vartheta + C\cos^2\vartheta. \qquad (1.3)$$

Each of the processes (2) is characterized by its own coefficients A, B, C, which are determined by the energy of the γ-quantum.

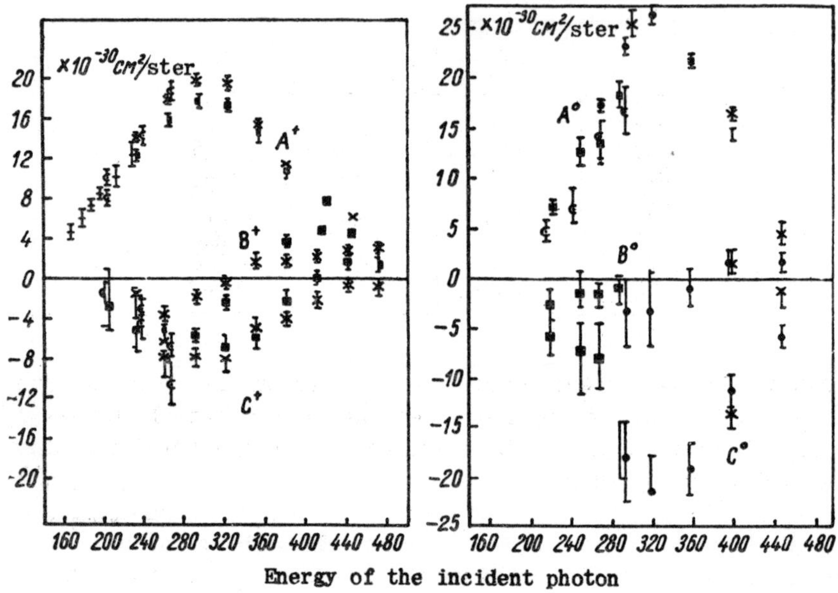

Fig. 1

The angular distribution of mesons formed by photons on protons has been studied by many experimenters, who have obtained results differing somewhat from one another. The "mean" values of the experimental coefficients A^+, B^+ and C^+ for the process (2a) and and the coefficients A^0, B^0, C^0 for the process (2b) are given (3) in Fig. 1[1]. The values of these coefficients are also given in Table 2(3). As can be seen, the isotropic terms A^+ and A^0 at first increase, reaching a maximum at $E_\gamma \approx 300$ Mev, and then decrease rapidly.

1 In this and all the following diagrams the energy E_γ in Mev in the laboratory system of co-ordinates is plotted along the abscissa.

TABLE 2

Mean values of the coefficients characterizing photo-production (10 cm^2/ster)

E_γ	A^+	B^+	C^+	A^0_{exp}	C^0_{exp}	A^0	C^0
200	9.3 ± 0.5	−1.8 ± 1.0	−3.3 ± 1.6	3.5 ± 0.5	− 3.3 ± 2.0	3.0 ± 0.4	− 2.8 ± 1.7
230	13.1 ± 0.5	−3.0 ± 0.9	−5.7 ± 1.0	7.4 ± 1.0	− 6.8 ± 2.0	6.6 ± 0.9	− 6.1 ± 1.8
260	16.9 ± 0.5	−4.3 ± 1.0	−7.4 ± 1.0	13.3 ± 1.5	−11.6 ± 4.0	12.2 ± 1.4	−10.7 ± 3.7
290	18.9 ± 0.5	−3.8 ± 1.2	−7.9 ± 0.9	22.0 ± 1.0	−17.6 ± 4.0	20.7 ± 0.9	−16.6 ± 3.8
320	18.4 ± 0.5	−1.1 ± 0.9	−7.2 ± 0.9	25.8 ± 1.0	−22.0 ± 4.0	24.6 ± 1.0	−20.9 ± 3.8
350	15.3 ± 0.5	+1.1 ± 0.6	−5.4 ± 0.9	23.3 ± 0.8	−20.8 ± 3.0	22.4 ± 0.8	−20.0 ± 2.9
380	11.3 ± 0.4	+2.5 ± 0.6	−3.0 ± 0.7	17.8 ± 1.5	−15.0 ± 2.5	17.2 ± 0.5	−14.5 ± 2.4
410	8.2 ± 0.4	+3.2 ± 0.8	−0.8 ± 0.8	12.8 ± 0.4	−10.0 ± 1.5	12.4 ± 0.4	− 9.7 ± 1.5
440	6.2 ± 0.3	+3.5 ± 0.7	+0.4 ± 0.8	9.2 ± 0.4	− 6.8 ± 1.4	9.0 ± 0.4	− 6.6 ± 1.4
470	4.7 ± 0.3	+3.7 ± 0.6	+1.0 ± 0.8	7.0 ± 0.4	− 5.0 ± 1.3	6.8 ± 0.4	− 4.9 ± 1.3

The coefficient B^+ is negative up to the energy $E_\gamma \approx 340$ Mev, when it changes sign. B^0 is a negligibly small quantity in the entire region of energies considered, so the values of B^0 are not given in Table 2.

For energies $E_\gamma \leqslant 200$ Mev $C^+ \ll A^+$. For $E_\gamma > 200$ Mev C^+ is negative, it increases rapidly in absolute value up to $E_\gamma \approx 300$ Mev, then drops rapidly, changing sign at $E_\gamma \approx 420$ Mev. For all energies the value of C^0 is approximately twice as large as the value of C^+. Moreover, $\frac{A^0}{C^0} = 1.22 \pm 0.1$ for all $E_\gamma \leqslant 400$ Mev (8) (it is true that in the region of low energies this ratio has been measured less accurately).

When determining A^0_{exp} and C^0_{exp} from experimental data it is necessary to take into consideration the difference in the thresholds of photo-production of π^0 and π^+-mesons, whereas in theoretical considerations this difference is not taken into account. For this reason, when comparing theory with experiment A^0_{exp} and C^0_{exp} have to be converted to A^0 and C^0, appropriate to identical thresholds of photo-production of π^0 and π^+-mesons. The calculated values of A^0 and C^0 are given in Table 2.

2. DIFFERENTIAL CROSS-SECTION OF THE PHOTO-PRODUCTION OF π-MESONS ON NUCLEONS

For the analysis of the experimental results on the photo-production of mesons on nucleons we shall find the expression for the meson angular distribution. We use for this purpose two hypotheses not open to doubt: the laws of conservation of angular momentum and parity[1].

The reaction considered may be schematically represented as follows[2]:

Photon {
- l – relative orbital angular momentum
- i – spin = 1
- m_i – projection of the spin
- L – total angular momentum
- m_L – projection of L
- p = 1 – electrical multipoles; 0 – magnetic multipoles

} + nucleon {
- I – spin
- m_I – projection of the spin

} →

---

1 It should be emphasized that only strong interactions, for which the law of conservation of parity holds, will be discussed below. (As is well-known, this law is not valid for weak interactions.)

2 The principal quantities characterizing the system and the notations for these used below are given in brackets. It should be noted that all calculations below are in the centre-of-mass system. The final state of a system is determined by the corresponding quantum numbers with a dash.

$$\rightarrow \quad \text{intermediate state} \quad \begin{Bmatrix} J - \text{total angular momentum} \\ M - \text{projection of } J \end{Bmatrix} \quad \rightarrow$$

$$\rightarrow \quad \begin{matrix} \text{recoil nucleon} \\ + \\ \text{meson} \end{matrix} \quad \begin{cases} l' - \text{relative orbital angular momentum of the meson} \\ \mu' - \text{projection of } l' \\ s' = \tfrac{1}{2} - \text{sum of the spins of the recoil nucleon and the meson} \\ m_s' - \text{projection of } s' \end{cases} \qquad (2.1)$$

When the photon is incident on a proton, for example, reactions like the following are possible:

$$\gamma + p = \begin{cases} p' + \gamma' \\ n + \pi^+ \\ p + \pi^0 \end{cases} \qquad (2.2)$$

Each pair of particles determines a corresponding channel of reaction. The reaction channel determined by γp is called the entry channel.

The reaction channels determined by $\gamma' p'$, $n\pi^+$, $p\pi^0$, are called exit channels.

We shall characterize the entry channel in reaction (2.1) by the total angular momentum L and the parity of the incident photon and also by the spin I of the nucleon. The state of the system after collision can be designated by the index of the channel α, characterizing the type of the particle formed (meson or γ-quantum), and in the case of the formation of a meson by the relative orbital angular momentum of the meson l' and the spin of the channel s', which is equal to the vector sum of the spins of the nucleon and the meson, i.e. $s' = \tfrac{1}{2}$.

We shall characterize the probability of transition of a system with the total angular momentum J from the channel αLs to the channel $\alpha' l' s'$ by the quantity $S^J_{\alpha' l' s';\, \alpha l s} \equiv S$.

A reaction with a single possible channel is characterized by the complex number S_{11}. If the incident wave for this channel is written in the form $\exp(-ikz)$, the outgoing wave will have the form $S_{11}\exp(ikr)$. If there is one channel, what goes in must come out, hence the modulus of S_{11} will be 1. Therefore S_{11} can be represented, in particular, in the form $\exp(2i\alpha)$, where α is the phase of the scattering.

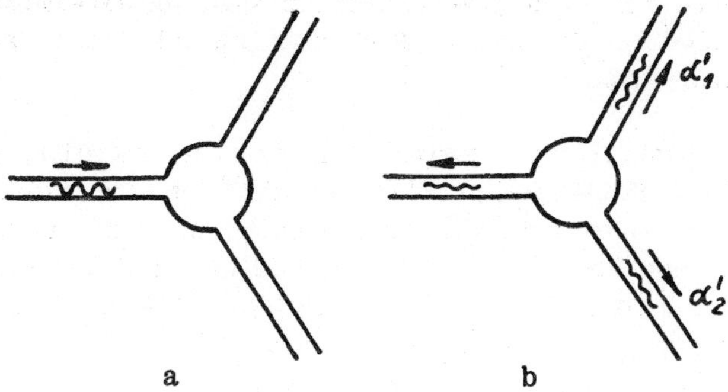

Fig. 2

When there are one entry and N exit reaction channels S_{ik} is a complex square matrix of order N (the scattering matrix). For example, for the process (2.2) which has three exit channels, the scattering matrix S_{ik} will be a square matrix with three rows. Its diagonal terms will be the probability amplitudes of the processes of elastic scattering: $p + \gamma \to p' + \gamma'$ (element S_{11}), $n + \pi^+ \to n' + \pi^{+\prime}$ (element S_{22}), $p + \pi^0 \to p' + \pi^{0\prime}$ (element S_{33}), and the non-diagonal terms the probability amplitudes for inelastic collisions of the type $p + \gamma \to n + \pi^+$ (element S_{12}), $\pi^+ + n \to p + \pi^0$ (element S_{23}) etc. Each of such processes will be characterized by its own values of α_{ik}.

It is useful to introduce the S-matrix because the expression for the differential cross-section of the scattering process (9, 10) can be written in a convenient form with its help.

The principle behind the derivation of this expression is as follows[1].

1 The considerations used below are similar to those used in the study of the scattering of a particle in a central field (11, 12), see also (13).

The reaction channels can be provisionally represented as shown in Fig. 2. Before the reaction (Fig. 2a) there is in the entry channel a plane incoming wave and in the exit channels the wave function is zero. After the reaction (Fig. 2b) outgoing waves will appear in all the channels. After the process is established therefore, both incoming and outgoing waves will be present in the entry channel and only outgoing waves in the exit channels. The cross-section of a given reaction will be determined by the square of the coefficient in the outgoing spherical wave of the corresponding channel.

We shall look for the expression for the outgoing spherical wave in the following way. First we shall write the expression (2.7) for the wave function of the entry channel α, which represents the superposition of the incoming (with the coefficient A) and the outgoing (with the coefficient B) waves:

$$\Phi_\alpha \approx A\Phi_\alpha^- + B\Phi_\alpha^+. \tag{2.3}$$

We then write a similar expression (2.9) for the exit channel

$$\Phi_{\alpha'} \approx D\Phi_{\alpha'}^+. \tag{2.3'}$$

The coefficients A, B and D are unknown quantities. The form of the coefficient A (2.12), and consequently of the coefficient D, can be found by comparing the coefficients in the incoming waves in expression (2.7) and the expression for the plane wave (2.11)

$$D_{\alpha'} \approx S_{\alpha\alpha'} A_\alpha . \tag{2.3'a}$$

The expression (2.12') for the outgoing spherical wave in the exit channel α' can be found in the same way; its coefficient will determine the cross-section of the process.

Let us now pass on to more detailed calculations (10).

1. Let us find the expression for the wave function $\Phi_{\alpha Lp}$ of the entry channel. The total angular momentum J of the initial photon-nucleon system is equal to the sum of the total angular momentum of the photon L and the spin of the nucleon I. Since the wave function of the photon with the given L and the spin

function of the nucleon are known separately, it is not difficult (appendix 1) to construct the asymptotic wave function of the entry channel, which describes the system photon-nucleon with the angular momentum J and its projection M:

$$\vec{\Phi}'^{M\,(\pm)}_{JLp} = \sum_{m_I} C^{JM}_{L,\,M-m_I,\,Im_I} \vec{A}^{M-m_I\,(\pm)}_{Lp} \chi^{m_I}_{I}, \tag{2.4}$$

where $C^{JM}_{j_1m_1,\,j_2m_2}$ are Clebsch-Gordan coefficients; $\chi^{m_I}_{I}$ is the spin wave function of the nucleon; $\vec{A}$ is the wave function of the photon.

The wave function of the photon with total angular momentum L and orbital momentum l is written as (appendix 3, formula (A. 13)):

$$\vec{A}^{m_L\,(\pm)}_{Lp} = -\sqrt{2}\sum_{l} C^{l0}_{L-1,\,11}\,\delta(l,p)\,i^{l-L-p}\,\vec{\Phi}^{m_L\,(\pm)}_{Ll}, \tag{2.5}$$

Here

$$\vec{\Phi}^{m_L\,(\pm)}_{Ll} = \sum_{m_l} C^{Lm_L}_{l,\,m_L-m_l,\,1m_l} Y^{m_L-m_l}_{l}(\vartheta)\vec{\chi}_{m_l} \times$$

$$\times \exp\left[\pm i\left(kr - \frac{\pi l}{2}\right)\right], \tag{2.6}$$

and

$\vec{\chi}_{m_l}$ is the spin function of the photon

$Y^{m}_{l}(\vartheta)$ – the normalized spherical harmonics

$C^{JM}_{j_1m_1,\,j_2m_2}$ – Clebsch-Gordan coefficients

k – the wave number of the photon

p – equal to 1 and 0 respectively for electric and magnetic multipoles. $\delta(l,1)$ and $\delta(l,0)$ differ from zero for $l = L \pm 1$ and $l = L$ respectively.

m_j – the projection of the corresponding angular momentum on the z-axis.

θ – the angle of scattering.

The most general form of the wave function for J and M in the entry channel will be a superposition of incoming and outgoing spherical waves with the spin-angle form (2.4), i.e.[1]

$$\vec{\Phi}_{\alpha Lp}(JM) = \frac{1}{r_\alpha \sqrt{c}} \left[A^{JM}_{\alpha Lp} \vec{\Phi}^{M(-)}_{JLp} + B^{JM}_{\alpha Lp} \vec{\Phi}^{M(+)}_{JLp} \right]. \qquad (2.7)$$

The unknown coefficients A and B characterize the incoming and outgoing waves respectively.

2. The total angular momentum J of the final system is equal to the sum of the relative orbital momentum of the meson l' and the spin of the channel s'. With the help of the wave function of the meson $Y^{m'}_{l'}$ and the channel spin function $\chi^{m}_{s'}{}^{s'}$, the asymptotic wave function of the exit channel with given J and M can be written as:

$$\psi^{M(\pm)}_{Jl's'} = \sum_{m_{s'}} C^{JM}_{l', M-m_{s'}, s'm_{s'}} Y^{M-m_{s'}}_{l'}(\vartheta)\, \chi^{m_{s'}}_{s'} \exp \times \times \left[\pm i \left(kr - \frac{\pi l'}{2} \right) \right]. \qquad (2.8)$$

The most general form of the wave function for J and M in the exit channel will be (if it is borne in mind that there are no incoming waves):

$$\Phi_{\alpha' l' s'}(JM) = \frac{1}{r_{\alpha'} \sqrt{v_{\alpha'}}} D^{JM}_{\alpha' l' s'} \psi^{M(+)}_{Jl's'}. \qquad (2.9)$$

The coefficient D characterizes the outgoing waves. We have

$$D^{JM}_{\alpha' l' s'} = \sum_{\alpha Lp} S_{\alpha' l' s',\, \alpha Lp} A^{JM}_{\alpha Lp}. \qquad (2.10)$$

3. The incident plane wave in the entry channel is defined as

$$\exp(ikz)\, \chi^{m}_{1} \chi^{m_I}_{I}. \qquad (2.10a)$$

---

1 Here and later the wave functions are normalized for unit current

$$v(\psi^* \psi)\, r = 1.$$

We consider the incident photon to move along the axis z, hence $m_i = \pm 1$. This means that there are only two transverse polarizations of the photon. The asymptotic expression for the plane wave resolved into the functions of (2.4) is (appendix 3, formula (A. 16)):

$$\exp(ikz)\,\vec{\chi}_{m_i}\chi_I^{m_I} = \frac{i}{kr}\sqrt{\frac{\pi}{2}}\sum_{LJp} i^{L+p}\,(2L+1)^{\frac{1}{2}}\,m^p \times$$

$$\times\, C^{JM}_{LmIm_I}\left[\vec{\Phi}'^{M(-)}_{JLp} - \vec{\Phi}'^{M(+)}_{JLp}\right] \qquad (2.11)$$

By comparing the coefficients of $\vec{\Phi}^{M(-)}_{JLp}$ in (2.7) and (2.11) we can find the amplitude of the incoming spherical waves

$$A^{JM}_{\alpha Lp} = \frac{i}{k_\alpha}\sqrt{\frac{\pi c}{2}}\, i^{L+p}\,\sqrt{2L+1}\,m^p C^{JM}_{Lm,\,Im_I}. \qquad (2.12)$$

In the other reaction channels $A^{JM}_{\alpha Lp} = 0$, because we assume that there is only one reaction channel.

From (2.3') it follows, that in order to determine the amplitude of photo-production it is necessary to find the coefficient in the outgoing spherical wave of the expression (2.9). On substituting (2.12) in (2.10) and the expression obtained in (2.9) we find

$$\Phi_{\alpha' l' s'} = \frac{i}{k_\alpha}\sqrt{\frac{c}{v_{\alpha'}}}\,\frac{\exp(ik_{\alpha'}r_{\alpha'})}{r_{\alpha'}}\,f_{\alpha\alpha'}\chi^{m_{s'}}_{s'}, \qquad (2.12')$$

where $f_{\alpha\alpha'}$ is the required amplitude of photo-production with a definite value of the projection for each of the particle spins

This is given by

$$f_{\alpha\alpha'} = \sum_{JLpl'} i^{L-l'+p}\sqrt{\frac{\pi}{2}}\,(2L+1)^{\frac{1}{2}}\,m^p C^{JM}_{Lm,\,Im_I} \times$$

$$\times\, C^{JM}_{l'\mu',\,s'm_{s'}} S^{J}_{\alpha' l' s',\,\alpha Lp} Y^{\mu'}_{l'}(\vartheta). \qquad (2.13)$$

The differential cross-section of photo-production averaged

over the initial projections (m and m_I) and summed over the final projections m_s' of the spin can be written as follows[1] with the help of

$$\sigma(\vartheta) = \frac{1}{4k^2} \sum_{m_{s'} m_I m} |f_{\alpha\alpha'}(\vartheta)|^2. \tag{2.14}$$

If, moreover, the formula (A. 6) expressing the product of two spherical functions in terms of the sum with respect to one spherical function is used, the expression for $\sigma(\theta)$ can be written as[2]

$$\sigma(\vartheta) = \frac{1}{4k^2} \sum_{\substack{J_1 J_2 L_1 L_2 \\ p_1 p_2 l_1' l_2' n}} i^{-L_1+L_2-p_1+p_2+l_1'-l_2'} S^*(1) S(2) \times$$

$$\times \left[(2L_1+1)(2L_2+1)(2l_1'+1)(2l_2'+1)\frac{1}{2n+1}\right]^{\frac{1}{2}} \times$$

$$\times C^{n0}_{l_1'0 l_2'0} \sum_{m_{s'} m_I m} (-1)^{\mu'} m^{p_1+p_2} C^{J_1 M}_{L_1 m, l m_1} C^{J_2 M}_{L_2 m, l m_1} C^{J_1 M}_{l_1' \mu', s' m_{s'}} C^{J_2 M}_{l_2' \mu', s' m_{s'}} \times \tag{2.15}$$

$$\times C^{n0}_{l_1' -\mu', l_2' \mu'} \frac{\sqrt{\pi}}{4} Y_{n0}(\vartheta).$$

The latter expression can be considerably simplified if Racah's formula (appendix 2) is used. With the help of this formula double summation over the projections of the moments $m_{s'}$ and m_I can be carried out in (2.15).

In order to find the sum over $m_{s'}$

$$\sum_{m_{s'}} C^{J_1 M}_{l_1' \mu', s' m_{s'}} C^{J_2 M}_{l_2' \mu', s' m_{s'}} C^{n0}_{l_1' -\mu', l_2' \mu'} \tag{2.16}$$

we convert it to the usual form (A. 11). To do this we make the

---

1 It should be borne in mind that in the given case $I = ½$ and m takes the values $\pm$ 1.

2 Because of the symmetry of the problem with respect to ϕ, we put $\phi = 0$.

following substitutions in (2.16): $s'm_{s'} \rightarrow a\alpha$, $l_2'\mu' \rightarrow b\beta$, $l_1'\mu' \rightarrow d\delta$, $J_2M \rightarrow c\gamma$, $n0 \rightarrow f, -\varphi$ $J_1M \rightarrow e\varepsilon$ and make use of the symmetry properties (A.4) of the Clebsch-Gordan coefficients.

The sum over m_I is calculated in the same way

$$\sum_{m_I} C^{n0}_{J_2M,\ J_1-M} C^{J_1M}_{L_1m,\ Im_I} C^{J_2M}_{L_2m,\ Im_I} \tag{2.17}$$

(the first coefficient in this sum was obtained by summation over m_s).

The final expression for the differential cross-section of photo-production of mesons on nucleons can then be written as[1]:

$$\begin{aligned}\sigma(\vartheta) = \frac{1}{16k^2}\sum(-1)\, S^*(1)\, S(2)\, i^{n-L_1+L_2-p_1+p_2} \times \\ \times [(2L_1+1)(2L_2+1)(2J_1+1)(2J_2+1)]^{\frac{1}{2}} C^{n0}_{L_1 1,\ L_2-1} \times \\ \times W\left(L_1J_1L_2J_2;\ \tfrac{1}{2},\ n\right) Z\left(l_1'J_1l_2'J_2;\ \tfrac{1}{2}\,n\right) P_n(\cos\vartheta),\end{aligned} \tag{2.18}$$

where

$$\begin{aligned}Z(l_1J_1l_2J_2;\ sn) = i^{n-l_1+l_2}[(2l_1+1)(2l_2+1)(2J_1+1)(2J_2+1)]^{\frac{1}{2}} \times \\ \times W(l_1J_1l_2J_2;\ sn)\, C^{n0}_{l_10l_20};\end{aligned} \tag{2.18a}$$

$W(l_1J_1l_2J_2;\ sn)$ are Racah's coefficients (appendix 2)

$P_n(\cos\theta)$ – Legendre's polynomials

L – total angular momentum of the photon

l' – orbital momentum of the meson

---

1 The summation over m in (2.15) can be carried out very simply if it is borne in mind that

$$C^{n0}_{L_1m,\ L_2-m} = (-1)^{L_1+L_2-n} C^{n0}_{L_1-m,\ L_2m}.$$

J – total angular momentum of the intermediate state.

Summation is carried out over J_1, J_2, L_1, L_2, p_1, p_2, l'_1, l'_2 and n.

In the case of photo-production of mesons on nucleons considered by us, we take $I = ½$, $s' = ½$.

In practical calculations the following expressions of (10) can be used instead of (2.18):

a) if $L_1 + L_2 + n$ is even

$$\sigma(\vartheta) = \frac{1}{16k^2}\sum(-1)\,S^*(1)\,S(2)\,[(2J_1+1)(2J_2+1)]^{\frac{1}{2}} \times \\ \times Z(L_1L_1L_2L_2;\ 1n)\,W\left(L_1J_1L_2J_2;\ \frac{1}{2}n\right) \times \\ \times Z\left(l'_1J_1l'_2J_2;\ \frac{1}{2}n\right) P_n(\cos\vartheta); \tag{2.19}$$

b) if $L_1 + L_2 + n$ is odd

$$\sigma(\vartheta) = \frac{1}{16k^2}\sum(-1)\,S^*(1)\,S(2)\,(-1)^{p_1}\sqrt{\frac{2L_2+1}{L_2}} \times \\ \times [(2J_1+1)(2J_2+1)]^{\frac{1}{2}} Z(L_1L_1L_2+1L_2;\ 1n) \times \\ \times W\left(L_1J_1L_2J_2;\ \frac{1}{2}n\right) Z\left(l'_1J_1l'_2J_2;\ \frac{1}{2}n\right) P_n(\cos\vartheta). \tag{2.19'}$$

It is important to note that with the help of the expressions (2.19) one can consider the cases when one intermediate state (i.e. $J_1 = J_2 = J$) is possible, as also the cases when two intermediate cases (i.e. $J_1 \neq J_2$) are possible. In the latter case interference effects may occur and these too can be taken into account with the help of these expressions.

The value of n, which occurs in (2.19), varies within the

following limits:

$$|l_1' - l_2'| \leqslant n \leqslant |l_1' + l_2'|. \tag{2.20}$$

Besides,

$$p_1 + p_2 + L_1 + L_2 - n \text{ should be even} \tag{2.21}$$

since

$$p_1 + p_2 + L_1 + L_2 + l_1' + l_2' \text{— is even}$$

and

$$l_1' + l_2' + n \text{— is even.}$$

The first condition following from the law of conservation of parity and the second from the fact that the coefficients $C^{n0}_{l_1'0,\, l_2'0}$ entering Z differs from zero only when $l_1' + l_2' + n$ is even.

The expressions for Racah's coefficients $W\,(l_1 J_1 l_2 J_2;\ sn)$ for example, are given in Table 3.

TABLE 3

Expressions for Racah coefficients $W\,(l_1 J_1 l_2 J_2;\ \tfrac{1}{2} n)$

	$l_2 = J_1 + \frac{1}{2}$	$l_1 = J_1 - \frac{1}{2}$
$l_2 = J_2 + \frac{1}{2}$	$(-1)^{J_1+J_2-n} \times \left[\frac{(J_1+J_2+n+2) \times (J_1+J_2-n+1)}{(2J_1+1)(2J_1+2) \times (2J_2+1)(2J_2+2)}\right]^{1/2}$	$(-1)^{J_1+J_2-n} \times \left[\frac{(n-J_1+J_2+1) \times (n+J_1-J_2)}{2J_1(2J_1+1) \times (2J_2+1) \times (2J_2+2)}\right]^{1/2}$
$l_2 = J_2 - \frac{1}{2}$	$(-1)^{J_1+J_2-n} \times \left[\frac{(n+J_1-J_2+1) \times (n-J_1+J_2)}{(2J_1+1)(2J_1+2) \times 2J_2(2J_2+1)}\right]^{1/2}$	$(-1)^{J_1+J_2-n-1} \times \left[\frac{(J_1+J_2+n+1) \times (J_1+J_2-n)}{2J_1(2J_1+1)2J_2 \times (2J_2+1)}\right]^{1/2}$

Let us see how formula (2.19) can be used for finding the angular distribution of photo-mesons in some particular cases.[1]

It is convenient for this purpose to use the tables of numerical values for Racah coefficients (24) and Z-coefficients (25) given in the appendices.

1. Let the mesons be formed only by the magnetic dipole γ photon (i.e. $p_1 = p_2 = 0$, $L_1 = L_2 = 1$). Since the parity of the nucleon is positive, the two intermediate states $J = 3/2 +$ or $J = \frac{1}{2} +$ are possible. If the interference phenomena are neglected, each of the two intermediate states can be considered separately. The parity of the meson is defined by the expression $(-1)^{l'+1}$, where l' is the orbital angular momentum of the meson. From the law of conservation of parity it follows that in the given case the orbital angular momentum of the meson can take only odd values ($l' = 1, 3, 5 \ldots$) while the law of conservation of angular momentum allows only $l' = l'_1 = l'_2 = l$, i.e. the magnetic dipole γ-quantum can form mesons only in the p-state. The quantity n can assume three values in this case: $n = 0, 1, 2$. The value $n = 1$ does not satisfy the condition $l'_1 + l'_2 + n$ is even and should therefore be rejected. The value $n = 2$ for $J = \frac{1}{2}$ also need not be considered. In this case $W(1\frac{1}{2}\ 1\frac{1}{2};\ \frac{1}{2}2) = 0$, since the condition $f \leqslant b + d$ is not fulfilled (see appendix 2).

Substituting the given values of the quantities in (2.19) we obtain[2]

a) for $J = 3/2 +$

$$\sigma(\vartheta) = \frac{1}{2}\left| M_d\left(\frac{3}{2}\right)\right|^2 \cdot (5 - 3\cos^2\vartheta), \tag{2.22}$$

where $M_d\,(3/2)$ is the transition matrix element characterizing the production of a meson in a p-state by magnetic dipole radiation

---

1 The expressions for $\sigma(\theta)$ given below can be obtained by other means also (14), (18).

2 Henceforth we shall omit the factor $1/_8\ k^2$ in expressions for the differential cross-section.

via the intermediate state with $J = 3/2$.

b) for $J = ½ +$

$$\sigma(\vartheta) = \left| M_d\left(\frac{1}{2}\right)\right|^2 = \text{const}, \tag{2.22'}$$

where M_d (½) is the transition matrix element characterizing the production of a meson in a p-state by magnetic dipole radiation via the intermediate state with $J = ½$.

2. Similarly in the case of electrical dipole radiation

a) for $J = ½ -$

$$\sigma(\vartheta) = \left| E_d\left(\frac{1}{2}\right)\right|^2 = \text{const}; \tag{2.23}$$

b) for $J = 3/2 -$

$$\sigma(\vartheta) = \left| E_d\left(\frac{3}{2}\right)\right|^2 \cdot \frac{3}{2}(1 + \cos^2\vartheta), \tag{2.24}$$

where E_d (½) and E_d (3/2) are the corresponding transition elements.

3. Since a number of intermediate states are possible in the same reaction, interference effects have to be taken into consideration. Thus for example, if two different intermediate states are possible: $J = 3/2 +$ and $J = ½ -$, then

$$\sigma(\vartheta) = |E_d|^2 - 2\mathrm{Re}\left[E_d^* M_d\left(\frac{3}{2}\right)\right]\cos\vartheta + $$
$$+ \frac{1}{2}\left| M_d\left(\frac{3}{2}\right)\right|^2 (5 - 3\cos^2\vartheta). \tag{2.25}$$

As can be seen an interference term appears in this case.

4. Finally, we shall find an expression for the angular distribution of photo-mesons which would allow analysis of the experi-

mental data obtained. We shall confine ourselves to a consideration of the case when mesons are formed in s- and p-states. In the s-state mesons can be formed by electric dipole radiation (we shall denote the corresponding matrix element by E_d) and in the p-state by magnetic dipole (matrix elements M_d (½) and M_d (3/2) and electric quadrupole radiations (matrix elements E_q). If the transitions enumerated and the interference between them are taken into consideration, the expression for the angular distribution of photo-mesons can be written as follows with the help of (2.19) [1];

$$\sigma(\vartheta) = |E_d|^2 + \left|M_d\left(\frac{1}{2}\right)\right|^2 + \frac{1}{2}\left|M_d\left(\frac{3}{2}\right)\right|^2 (2 + 3\sin^2\vartheta) +$$

$$+ \frac{1}{8}|E_q|^2(1 + \cos^2\vartheta) - 2\mathrm{Re}\left[E_d^*\left(M_d\left(\frac{3}{2}\right) - M_d\left(\frac{1}{2}\right) - \frac{1}{2}E_q\right)\right] \times$$

$$\times \cos\vartheta - \frac{1}{2}\mathrm{Re}\left[E_q^*\left(M_d\left(\frac{3}{2}\right) - M_d\left(\frac{1}{2}\right)\right)\right](3\cos^2\vartheta - 1) -$$

$$- \mathrm{Re}\left[M_d^*\left(\frac{3}{2}\right) M_d\left(\frac{1}{2}\right)\right](3\cos^2\vartheta - 1). \tag{2.26}$$

The expression for $\alpha(\theta)$ for the case when mesons are formed in s-, p- and d- states can also be found without difficulty with the help of (2.19) (15).

We shall now show that the arguments of the matrix elements will be the corresponding scattering phases for mesons on nucleons, i.e. known quantities.

---

1. Here we have changed over to the following new notations for the unknown matrix elements: $E_d \to E_d$, $M_d\left(\frac{1}{2}\right) \to M_d\left(\frac{1}{2}\right)$, $M_d\left(\frac{3}{2}\right) \to M_d\left(\frac{3}{2}\right)$, $E_q \to \frac{1}{2\sqrt{3}} E_q$; The expression for $\sigma(\vartheta)$ is obtained in the same way as in paper (3).

3. RELATIONSHIP BETWEEN THE PROCESSES OF SCATTERING AND PHOTO-PRODUCTION

If we confine ourselves to the energy range in which mesons are formed only in *s*- and *p*-states, the angular distribution of the mesons will be described by the expression (2.26). Each of the four processes of photo-production (2) is characterized by its coefficients E_d, M_d (½), M_d (3/2), E_q. Consequently there will be 16 unknown complex coefficients in all.

It is more convenient, however, to classify the final state of the meson – nucleon system not according to its charge, but according to the total isotopic spin, which in this case can take the values 3/2 and ½. Then, by the use of the hypothesis of charge symmetry, each of the coefficients E_d, M_d (½), M_d (3/2), E_q for all the four processes (2) can be expressed by three quantities, since instead of 16 unknown coefficients there will be 12.

This can be done as follows (16).

We shall start from Schrödinger's equation which describes the meson-nucleon system;

$$i\frac{\partial\psi}{\partial t} = \hat{H}_0\psi, \tag{3.1}$$

where $\hat{H}_0$ is the Hamiltonian of the meson-nucleon system.

In order to introduce in the Hamiltonian the electromagnetic field which determines the processes (2), $\frac{\partial}{\partial x_i}$ and $\frac{\partial}{\partial X_i}$, operating on the nucleon and meson functions have to be replaced by the following expressions respectively:

$$\frac{\partial}{\partial x_i} \rightarrow \frac{\partial}{\partial x_i} + ie\hat{q}A_i \; ;$$

$$\frac{\partial}{\partial X_i} \rightarrow \frac{\partial}{\partial X_i} + ie\hat{L}_3 A_i, \tag{3.2}$$

where $\hat{L}_3$ is the meson charge operator;

$\hat{q} = \frac{1+\hat{\tau}_3}{2}$ – the nucleon charge operator;

$\hat{\tau}_3$ – the projection of the isotopic spin operator of the nucleon on the z-axis.

The electromagnetic interaction can be considered as a small perturbation. The Hamiltonian obtained after the substitution (3.2) can then be expanded in a series of powers of eA_i. If we confine ourselves to terms of the first degree smallness, we shall obtain the following instead of equation (3.1);

$$i\frac{\partial\psi}{\partial t} = (\hat{H}_0 + \hat{H}_e)\,\psi. \tag{3.3}$$

Here

$$\hat{H}_e = e\hat{a}\hat{q}A_i + e\hat{b}\hat{L}_3 A_i + \ldots, \tag{3.3'}$$

where $\hat{a}$ and $\hat{b}$ are expansion coefficients containing $\hat{H}_0$.

The transition matrix element of the meson – nucleon system from the initial state (in our case nucleon) ψ_0 to the final state (meson – nucleon) ψ_f can be written as

$$T = (\psi_f \hat{H}_e \psi_0). \tag{3.4}$$

Now let us make use of the hypothesis of charge symmetry which states that the Hamiltonian $\hat{H}_0$ is invariant with respect to a rotation of 180^0 about the x-axis in isotopic space (i.e. the meson – nucleon system is invariant with respect to the substitution $\pi^{\pm} \rightleftarrows \pi^{\mp}$ and neutron $\rightleftarrows$ proton). If the rotation mentioned is executed, $\hat{a}$ and $\hat{b}$ in (3.3') will remain unchanged and

$\hat{\tau}_3$ and $\hat{L}_3$ will change signs, so that (omitting the factor eA_i)

$$\hat{H}'_e = \hat{a}\frac{1-\hat{\tau}_3}{2} + \hat{b}\,(-\hat{L}_3) = \frac{1}{2}\hat{a} - \left(\hat{a}\frac{\hat{\tau}_3}{2} + \hat{b}\hat{L}_3\right). \quad (3.4a)$$

Hence $\hat{H}_e$ can be resolved into two terms for a rotation through 180^0 about the x-axis in isotopic space

$$\hat{H}_e = \hat{S}_0 + \hat{V}_3, \quad (3.5)$$

One of these, $\hat{S}_0 = ½\,\hat{a}$ does not change sign, while the other, $\hat{V}_3$ does, i.e. behaves like the third component of a vector with respect to the rotation.

If (3.5) is taken into account the matrix element (3.4) can be rewritten as

$$T = (\psi_f\,|\,\hat{S}_0 + \hat{V}_3\,|\,\psi_0). \quad (3.6)$$

If ψ is expressed in the form of the product of a function of isotopic spin Φ and the function ψ' of the remaining variables, and $\hat{S}_0$ and $\hat{V}_3$ in the form of the product of operators of isotopic spin $\hat{S}''_0$, $\hat{V}''_3$ and the operators of the remaining quantities $\hat{S}'_0$, $\hat{V}'_3$, then

$$T = (\psi'_f\Phi_f\,|\,\hat{S}''_0\hat{S}'_0 + \hat{V}''_3\hat{V}'_3\,|\,\psi'_0\Phi_0). \quad (3.6')$$

As an example let us calculate the transition matrix element for the process $\gamma + p \rightarrow n + \pi^+$. In this case the isotopic function of the final system $n\pi^+$ has the form[1]

$$\Phi_f = \sqrt{\frac{1}{3}}\,\Phi_{3/2,\,-1/2} - \sqrt{\frac{2}{3}}\,\Phi_{1/2,\,-1/2}. \quad (3.7)$$

The initial state is characterized by the isotopic function

$$\Phi_0 = \varphi_{1/2,\,-1/2}. \quad (3.7')$$

1. As in the paper (3), for the proton and neutron respectively we take $I_z = -½$, and $I_z = +½$ for the π^+-, π^--, π^0-mesons $I_z = -1, +1, 0$.

Substituting (3.7) and (3.7') and (3.6') we get[1]

$$T^{+} = \sqrt{2}\, t_3 + \frac{1}{\sqrt{2}}\, t_1 - \sqrt{2}\, S, \tag{3.8}$$

where S and t_1 are the transition matrix elements, characterizing the production of a meson-nucleon system with isotopic spin ½, and t_3 with isotopic spin 3/2.

Similarly we find

$$T^{-} = \sqrt{2}\, t_3 + \frac{1}{\sqrt{2}}\, t_1 + \sqrt{2}\, S \quad (\gamma n \to p\pi^{-}), \tag{3.9a}$$

$$T^{0} = 2t_3 - \frac{1}{2}\, t_1 + S \quad (\gamma p \to p\pi^{0}), \tag{3.9b}$$

$$T^{n0} = 2t_3 - \frac{1}{2}\, t_1 - S \quad (\gamma n \to n\pi^{0}). \tag{3.9c}$$

Since the final state with a given isotopic spin can be formed by various multipoles of radiation, we shall introduce the following notation. Let $t_1\,(J,\, l',\, \phi)$ and $t_3\,(J,\, l',\, \phi)$ characterize the formation of a meson-nucleon system with the corresponding isotopic spin, total angular momentum J and orbital angular momentum l' by the multipole ϕ. Then the coefficient E_d^{+}, for example, corresponding to the formation of a π^{+}meson by an electric dipole radiation e_1 in the state $l = 0$, can be written as

$$E_d^{+} = \sqrt{2}\, t_3\,(1/2,\, 0,\, e_1) + \frac{1}{\sqrt{2}}\, t_1\,(1/2,\, 0,\, e_1) - \sqrt{2}\, S\,(1/2,\, 0,\, e_1). \tag{3.10}$$

---

1. Here it should be remembered that (see (12), p. 116) $\left(n' I' I_z' \middle| \hat{S}_0'' \hat{S}_0' \middle| n I I_z\right) = \delta_{II'} \delta_{I_z I_z} \left(n' \middle| \hat{S}_0' \middle| n\right) = \delta_{II'} \delta_{I_z I_z'} S;\ \left(n' \tfrac{1}{2} I_z' \middle| \hat{V}_3' \middle| n\, \tfrac{1}{2} I_z\right) = \delta_{I_z I_z'} I_z \left(n' \middle| \hat{V}_3' \middle| n\right) =$

$$= \delta_{I_z I_z'} I_z \cdot t_1;\ \left(n'\, \tfrac{3}{2} I_z' \middle| \hat{V}_3' \middle| n\, \tfrac{1}{2} I_z\right) = \delta_{I_z I_z'} \left[9/4 - I_z^2\right]^{1/2} \left(n' \middle| \hat{V}_{33}' \middle| n\right) =$$

$$= \delta_{I_z I_z'} \left[\frac{9}{4} - I_z^2\right]^{1/2} t_3.$$

n denotes the set of all the indices (besides the isotopic) which characterize the state of the system.

according to (3.8).

Similarly when a π^0-meson is formed, according to (3.9b)

$$E_d^0 = 2t_3\,(^1/_2,\ 0,\ e_1) - \frac{1}{2}\,t_1\,(^1/_2,\ 0,\ e_1) + S\,(^1/_2,\ 0,\ e_1). \quad (3.11)$$

As can be seen from (3.9), the coefficients E_d^- and E_d^{n0} are obtained from E_d^+ and E_d^0 respectively by changing the sign of S. The coefficients M_d (½), $M_d(3/2)$, E_q have the same form, but differ in the values of J, l, ϕ; e.g. for M_d (3/2): $J = 3/2$, $l = 1$, $\phi = m_1$.

Thus each of the coefficients in (2.26), for all the four processes (2), can be expressed by means of the three quantities S, t_1, t_3, so that the processes (2) will be determined by 12 complex coefficients S, t_1 and t_3.

We shall show (18, 19), that the arguments of t and S are the the corresponding phases of meson - nucleon scattering, i.e. each of the quantities t and S can be expressed in the form

$$e^{i\alpha}\cdot N, \quad (3.12)$$

where N is a real number,

α - the phase of the meson - nucleon scattering.[1]

To do this we have to make use of the fact that the scattering matrix is unitary:

$$S^+S = SS^+ = 1 \quad (3.13)$$

and symmetrical with respect to transition from the initial state to the final and from the final to the initial[2]:

$$S_{ik} = S_{ki}\,. \quad (3.14)$$

1. Of course all t and S are determined correctly except for a phase factor which is of no significance.

2. For details on this see (17).

Let there be three channels of reaction not connected with each other. Then the scattering matrix can be written in the following form:

$$S_0 = \begin{vmatrix} e^{2i\alpha_s} & 0 & 0 \\ 0 & e^{2i\alpha_3} & 0 \\ 0 & 0 & e^{2i\alpha_1} \end{vmatrix}, \qquad (3.15)$$

where α_s, α_1, α_3 are the corresponding phases for each channel.

If there is interaction between the channels[1],

$$S = S_0 + i\varepsilon. \qquad (3.16)$$

The matrix ϵ contains all the interaction terms (interaction between the channels). Let us assume that the elements ϵ are so small that their squares can be neglected.

Bearing (3.13) in mind, we find

$$SS^+ = S_0S_0^+ + i\varepsilon S_0^+ - iS_0\varepsilon^+. \qquad (3.17)$$

Since $S_0\ S_0^+ = 1$, we get from the last relationship

$$S_0\varepsilon^+ = \varepsilon S_0^+. \qquad (3.18)$$

With the help of (3.15) the expression (3.18) can be rewritten as[2]

$$\exp 2i\alpha_m\varepsilon^*_{nm} = \varepsilon_{mn}\exp(-2i\alpha_n). \qquad (3.19)$$

From the symmetry property of the S-matrix (3.14) it follows that

$$\varepsilon_{nm} = \varepsilon_{mn}. \qquad (3.19a)$$

Then the equation (3.19) can be rewritten as follows:

---

1. The factor i in ϵ is taken for the sake of convenience.
2. ϵ * denotes a complex-conjugate quantity.

$$\exp 2i\alpha_m \varepsilon^*_{nm} = \varepsilon_{nm} \exp(-2i\alpha_n), \tag{3.20}$$

or after multiplying both sides by $\exp\ [i\,(\alpha_n - \alpha_m)]$

$$\varepsilon_{nm} \exp[-i\,(\alpha_m + \alpha_n)] = \{\varepsilon_{nm} \exp[-i\,(\alpha_m + \alpha_n)]\}^*. \tag{3.21}$$

Hence $\varepsilon_{nm} \exp\ [-i\,(\alpha_m + \alpha_n)] = \rho_{nm}$ is a real quantity, in other words

$$\varepsilon_{nm} = \rho_{nm} \exp[i\,(\alpha_m + \alpha_n)], \tag{3.22}$$

i.e. each element of the S-matrix can actually be reduced to the form (3.12).

Let us use the result obtained, taking into account the reaction (2). If we confine ourselves to the case of the production of mesons only in the s-states, α_s in the expression (3.15) for the S-matrix will correspond to the phase of scattering of the photons on protons, and α_3 and α_1 to the phases of scattering of mesons on nucleons in the state with the isotopic spin $I = 3/2$ and ½. The non-diagonal matrix elements, corresponding to photo-production of the meson-nucleon system with $I = 3/2$ and ½, will be considerably smaller than the diagonal matrix elements as the cross-section of scattering is considerably larger than that of photo-production. Also, the cross-section of scattering of photons on protons is considerably smaller than the cross-section of photo-production of mesons so that we can write $\alpha_s = 0$. Hence the matrix for the case considered will appear as follows.

$$\begin{vmatrix} 1 & \varepsilon_{12} e^{i\alpha_3} & \varepsilon_{13} e^{i\alpha_1} \\ \varepsilon_{21} e^{i\alpha_3} & e^{2i\alpha_3} & 0 \\ \varepsilon_{31} e^{i\alpha_1} & 0 & e^{2i\alpha_1} \end{vmatrix}. \tag{3.22a}$$

ϵ_{12} should be taken as the modulus of the quantity t_3 and ϵ_{13} the modulus of the quantities t_1 and S. In other words

$$t_3\left(\frac{1}{2},\ 0,\ e_1\right) = e^{i\alpha_3} E_3;$$

$$t_1\left(\frac{1}{2},\ 0,\ e_1\right) = e^{i\alpha_1} E_1; \tag{3.23}$$

$$S\left(\frac{1}{2},\ 0,\ e_1\right) = \frac{1}{2} e^{i\alpha_1} E'_1\,, \tag{3.23}$$

where E_3, E_1, E'_1 are moduli of the quantities t_3, t_1, S.

Similarly, for the case of formation of mesons in the p-state:

$$t_3\,(^3/_2,\ 1,\ m_1) = e^{i\alpha_{33}} M_{33}\,; \qquad t_3\,(^3/_2,\ 1,\ e_2) = e^{i\alpha_{33}} E_{33}\,;$$
$$t_1\,(^3/_2,\ 1,\ m_1) = e^{i\alpha_{13}} M_{13}\,; \qquad t_1\,(^3/_2,\ 1,\ e_2) = e^{i\alpha_{13}} E_{13}\,;$$
$$S\,(^3/_2,\ 1,\ m_1) = \frac{1}{2} e^{i\alpha_{13}} M'_{13}\,; \qquad S\,(^3/_2,\ 1,\ e_2) = \frac{1}{2} e^{i\alpha_{13}} E'_{13};$$

$$\begin{aligned} t_3\,(^1/_2,\ 1,\ m_1) &= e^{i\alpha_{31}} M_{31};\\ t_1\,(^1/_2,\ 1,\ m_1) &= e^{i\alpha_{11}} M_{11};\\ S\,(^1/_2,\ 1,\ m_1) &= \frac{1}{2} e^{i\alpha_{11}} M'_{11}, \end{aligned} \tag{3.23'}$$

where M_{33}, M_{13} ... are moduli of the corresponding amplitudes.
α_{33}, α_{31} ... phases of scattering of mesons on nucleons.

The phases of scattering α_{ik} entering these expressions correspond to the energy (in the CM system) which the meson possesses in the final state in the formation process.

Substituting (3.23) in (3.10), (3.11) etc. we find

$$\left.\begin{aligned} E_d^+ &= \sqrt{2}\, e^{i\alpha_3} E_3 + \frac{1}{\sqrt{2}} e^{i\alpha_1} \left(E_1 - E'_1\right);\\ M_d^+\,(^1/_2) &= \sqrt{2}\, e^{i\alpha_{31}} M_{31} + \frac{1}{\sqrt{2}} e^{i\alpha_{11}} \left(M_{11} - M'_{11}\right);\\ M_d^+\,(^3/_2) &= \sqrt{2}\, e^{i\alpha_{33}} M_{33} + \frac{1}{\sqrt{2}} e^{i\alpha_{13}} \left(M_{13} - M'_{13}\right);\\ E_q^+ &= \sqrt{2}\, e^{i\alpha_{33}} E_{33} + \frac{1}{\sqrt{2}} e^{i\alpha_{13}} \left(E_{13} - E'_{13}\right) \end{aligned}\right\} \tag{3.24}$$

and

$$\left.\begin{aligned} E_d^0 &= 2e^{i\alpha_3}E_3 - \frac{1}{2} e^{i\alpha_1} \left(E_1 - E_1'\right); \\ M_d^0\,(^1/_2) &= 2e^{i\alpha_{31}}M_{31} - \frac{1}{2} e^{i\alpha_{11}} \left(M_{11} - M_{11}'\right); \\ M_d^0\,(^3/_2) &= 2e^{i\alpha_{33}}M_{33} - \frac{1}{2} e^{i\alpha_{13}} \left(M_{13} - M_{13}'\right); \\ E_q^0 &= 2e^{i\alpha_{33}}E_{33} - \frac{1}{2} e^{i\alpha_{13}} \left(E_{13} - E_{13}'\right). \end{aligned}\right\} \qquad (3.25)$$

In the same way the transition amplitudes for the different multipoles are expressed by the amplitudes of formation of the meson - nucleon system with a definite isotopic spin.

The values of the coefficients for E_d^-, $M_d^-\,(^1/_2)$, $M_d^-(^3/_2)$, E_q^-, as can be seen from (3.8) and (3.9a), are obtained from the formulae (3.24) by changing the sign of E_1', M_{11}', M_{13}', E_{13}'. From the expressions (3.9b) and (3.9c) it can be seen that to obtain the matrix elements E_d^{n0}, $M_d^{n0}\,(^1/_2)$, $M_d^{n0}\,(^3/_2)$, and E_q^{n0} it is necessary to change the signs of E_1', M_{11}', M_{13}', E_{13}' in the formulae (3.25).

Substituting (3.24) and expressions similar to them in (2.26), we shall obtain an expression in which the unknown quantities will be E and M. The experimental data on the angular distribution of mesons formed by γ-quanta can be used to determine these quantities. Comparison of (2.26) and (1.3) leads to the following relationship between E and M and the known experimental quantities A, B and C:

$$\begin{aligned} A &= |E_d|^2 + |X|^2 + |Y|^2; \\ B &= -2\mathrm{Re}\left(E_d^* K\right); \\ C &= |K|^2 - |X|^2 - |Y|^2, \end{aligned} \qquad (3.26)$$

where

$$X = {}^3/_2 M_d\,({}^3/_2) + \frac{1}{4} E_q;$$

$$Y = \frac{1}{2}\left[M_d\,({}^3/_2) - \frac{1}{2} E_q\right] + M_d\,({}^1/_2);$$

$$K = \left[M_d\,({}^3/_2) - \frac{1}{2} E_q\right] - M_d\,({}^1/_2). \tag{3.27}$$

The expressions (3.26) are very convenient for the analysis of experimental data.

4. ANALYSIS OF EXPERIMENTAL DATA

1. *The relative role of different multipoles in the photo-production of mesons* (3). From expressions (3.24) and (3.25) it can be seen that the differential cross-sections of photo-production of mesons are determined by the set of values of E and M. It is interesting to find out what is the relative contribution to the cross-section of each of these quantities, particularly of those which correspond to the formation of the meson – nucleon system with $J = I = 3/2$ in the final state.

CASE 1. Let us first consider the simplest case, assuming that besides E_d, only M_{33} plays a fundamental role in processes of photo-production.

In this case the expressions (3.26) will take the following form:[1]

$$\left.\begin{aligned}
A^+ &= A_s + A_p^+ = A_s + {}^5/_2 A_M;\\
C^+ &= -{}^3/_2 A_M;\\
B^+ &= -{}^3/_2 (A_s A_M)^{1/2} [\cos(\alpha_{33} - \alpha_3) + 2\cos(\alpha_{33} - \alpha_1)];\\
A^0 &= 2A_p^+;\\
C^0 &= 2C^+,
\end{aligned}\right\} \quad (4.1)$$

where

$$\begin{aligned}
A_s &= |E_d^+|^2;\\
A_p^+ &= A^+ - A_s;\\
A_M &= 2M_{33}^2.
\end{aligned} \quad (4.1a)$$

---

1 In this and later cases we neglect the contribution of s-wave to the cross-section $\sigma^0(\theta)$, i.e. we put $E_d^0 = 0$, and hence, $B^0 = 0$, because this contribution is comparatively small, according to experimental data.

The quantity A_s represents the contribution of the s-wave to $\sigma^+(\theta)$.

The equations (4.1) connect five known quantities A^+, B^+ etc. with two unknowns A_s and A_M, i.e. the system of equations (4.1) is over-determined.

Since the experimental data on the angular distribution of π^+-mesons are more accurate than those for π^0-mesons, it is reasonable to use the first two equations of the system (4.1), for example, for the determination of the two unknown quantities. The quantities A_s and A_M can be determined with the help of A^+ and C^+ by making use of the scattering phases given in Table 1. B^+ can then be calculated from the third equation of the system (4.1). The calculated and experimental values for B^+ are given in Fig. 3. As can be seen, there is a considerable difference between these values.

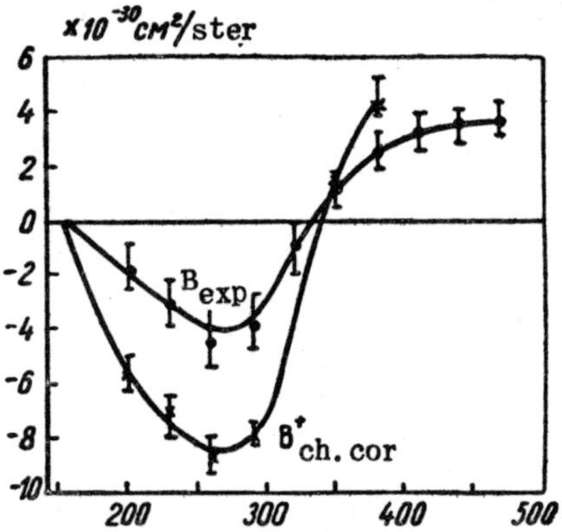

Fig. 3

Further, it follows from the equations (4.1) that for all energies $A^0/C^0 \simeq 1.67$. From experimental data it is found that $A^0/C^0 = 1.22 \pm 0.1$, i.e. a different value. (It is true that for small energies this ratio has not been measured very accurately).

To explain the experimental data therefore, it is not enough to confine oneself (together with E_d) to only one matrix element, even if it is the most important matrix element M_{33}.

CASE 2. If all the quantities corresponding to non-resonant states (e.g. M_{13} etc.) are neglected and all the products of the moduli of magnetic dipole and electric quadrupole amplitudes are retained (in addition to E_d), the equations (3.26) will take the form

$$\left.\begin{aligned} A^+ &= A_s + A_p^+ = A_s + A_{\varkappa 0} + \frac{1}{4} A_{k0};\\ C^+ &= {}^3/_4 A_{k0} - A_{\varkappa 0};\\ B^+ &= -\frac{2}{3}(A_s A_{k0})^{1/2}[\cos(\alpha_{33} - \alpha_3) + 2\cos(\alpha_{33} - \alpha_1)];\\ A^0 &= 2A_p^+,\\ C^0 &= 2C^+, \end{aligned}\right\} \quad (4.2)$$

where

$$A_{k0} = 2\left(M_{33} - \frac{1}{2}E_{33}\right)^2; \quad (4.2a)$$

$$A_{\varkappa 0} = \frac{1}{2}\left(3M_{33} + \frac{1}{2}E_{33}\right)^2. \quad (4.2b)$$

The quantity $A_{\varkappa 0}$ represents the contribution of the resonance p-wave which does not interfere and A_{k0} the one that interferes with the s-wave.

If the quantity E_{33} is neglected in the equations (4.2), we get (4.1).

From the equations (4.2a and 2b) it can be seen that each of the quantities M_{33} and E_{33} can be determined separately with the help of A_{k0} and $A_{\varkappa 0}$. In equations (4.2) there are three unknowns: A_s, $A_{\varkappa 0}$ and A_{k0}. Determining these from the first three equations of the system (4.2) and substituting them in the two remaining equations, we calculate A^0 and C^0. The calculated (dotted curves) and experimental values for A^0 and C^0 are given in Fig. 4. As can be seen, for each of these quantities the discrepancy is greater than that caused by experimental errors. Consequently, to explain the experimental data it is not enough to take into account (in addition to E_d) only the resonance terms and the products of the moduli of magnetic dipole and electric quadrupole amplitudes.

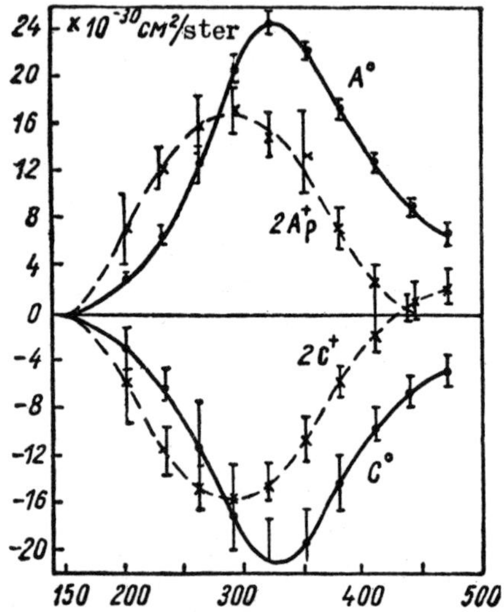

Energy of the incident photon

Fig. 4

CASE 3. Let us calculate (in addition to E_d) the contribution of the non-resonant p-waves. For this only the squares of the non-resonance terms should be neglected. The expressions (3.26) can then be rewritten as follows:[1]

$$\left.\begin{aligned}
A^{+} &= A_s + A_p^{+} = A_s + A_{x0} + \frac{1}{4} A_{k0} + A_{\Delta_1} \cos \alpha_{33};\\
B^{+} &= -{}^{2}/_{3} (A_s A_{k0})^{1/2} [\cos(\alpha_{33} - \alpha_3) + 2 \cos(\alpha_{33} - \alpha_1)];\\
C^{+} &= \frac{3}{4} A_{k0} - A_{x0} - A_{\Delta_1} \cos \alpha_{33};\\
A^{0} &= 2A_p^{+} - 3A_{\Delta_1} \cos \alpha_{33};\\
C^{0} &= 2C^{+} + 3A_{\Delta_1} \cos \alpha_{33},
\end{aligned}\right\} \qquad (4.3)$$

1 Strictly speaking, in deriving the equations (4.3) we have neglected two other quantities less than A_{Δ_1} (see (3) formulae (8.7)).

where

$$A_{\Delta_1} = \frac{1}{\sqrt{2}} X_0 \left\{3\left[M_{13} - M'_{13}\right] + \frac{1}{2}\left[E_{13} - E'_{13}\right]\right\} +$$

$$+ \frac{1}{2\sqrt{2}} K_0 \left\{\left[M_{13} - M'_{13}\right] - \frac{1}{2}\left[E_{13} - E'_{13}\right] + 2\left[M_{11} - M'_{11}\right]\right\},$$

$$X_0^2 = A_{x0}; \qquad K_0^2 = A_{\kappa 0}. \tag{4.3a}$$

The quantity A_{Δ_1} represents the contribution of the non-resonance p-waves. It is interesting to note that A_{Δ_1} enters only in the form of the product $A_{\Delta_1} \cdot \cos\alpha_{33}$. Since $\cos \alpha_{33}$ passes through zero for $E\gamma \approx 345$ Mev (Table 1), the part played by the non-resonance p-waves in this range of energy must be small.

The unknowns in the equations (4.3) are A_s, A_{x0}, A_{k0}, A_{Δ_1}.

From equations (4.3) it can be seen that

$$A^+ + C^+ = A_s + A_{k0}. \tag{4.4}$$

From this equation, and also from the equation for B^+ it is possible to determine the values of A_s and A_{k0}, i.e. the contribution of the s- and p-waves to $\sigma^+(\theta)$ can be determined separately by using the angular distribution of the π^+-mesons.

To determine $A_p^+ = A_{x0} + A_{\Delta_1}\cos\alpha_{33} + \frac{1}{4}A_{k0}$

we use the equation

$$A_p^+ = A^+ - A_s. \tag{4.5}$$

When A_p^+ and C^+ are known, the value of $A_{\Delta_1}\cos\alpha_{33}$ can be determined with the help of either of the two equations of the system (4.3). Such an independent determination gives for $A_{\Delta_1} \cos \alpha_{33}$ values close to each other, as can be seen from a comparison of the figures in the last two columns of Table 4. The values of A_s, A_{k0}, A_p^+ and A_{x0} found from the equations (4.3) for different energies are given in the same table.

TABLE 4

Table of values for A_S, A_{k0}, A_p^+, A_{x0}, $A_{\Delta_1}\cos\alpha_{33}$ (10^{-30} cm^2/ster)

E_γ	A_s	A_{k0}	A_p^+	A_{x0}	$A_{\Delta_1}\cos\alpha_{33}$	$A_{\Delta_1}\cos\alpha_{33}$
200	5.9±1.5	0.14±0.15	3.4±1.6	2.1±1.8	1.3±1.1	1.3±1.2
230	7.1±0.9	0.33±0.20	6.0±1.0	4.1±1.2	1.8±0.7	1.8±0.9
260	8.9±0.9	0.61±0.29	8.0±1.0	6.6±1.2	1.3±0.8	1.4±1.3
290	10.4±0.8	0.65±0.41	8.5±0.9	9.3±1.2	—1.2±0.7	—0.3±1.3
320	10.9±0.8	0.24±0.40	7.5±0.9	10.4±1.2	—3.3±0.7	—2.2±1.4
350	8.5±1.6	1.38±1.50	6.8±1.7	9.5±2.0	—2.9±1.2	—3.1±1.2
380	7.8±0.7	0.55±0.26	3.5±0.8	6.6±1.0	—3.4±0.6	—2.8±0.9
410	6.9±0.6	0.44±0.22	1.3±0.8	4.1±0.9	—3.3±0.6	—2.7±0.7
440	6.1±0.6	0.54±0.22	0.1±0.7	2.7±0.9	—2.9±0.9	—2.5±0.7
470	5.0±0.6	0.69±0.22	—0.3±0.7	1.9±0.9	—2.5±0.9	—2.3±0.7

The values of $A_{\Delta_1}\cos\alpha_{33}$ are depicted graphically (continuous curve) in Fig. 5. From the graph it can be seen that cos α_{33} becomes zero when $E\gamma\approx 275$ Mev. This contradicts the experimental data, according to which cos α_{33} passes through zero when $E\gamma\approx 345$ Mev (Table 1). There are two ways of removing this contradiction (besides the assumption that the resonance model is incorrect). Firstly, it may be thought that the scattering phases α_{31}, α_{13}, α_{11} omitted as small are not really small. The contradiction may in fact be removed by taking these phases into account, but for this it has to be assumed that these phases become sufficiently large in magnitude, for example, larger than 30° for $E\gamma\approx 300$ Mev. Such large values for the phases α_{11}, α_{31}, α_{13} go against the phase analysis of the experimental data on the scattering of mesons (1). This explanation of the divergence between theory and experiment is therefore not very probable.

The second way of explaining the discrepancy amounts to taking into account the higher scattering phases (d-phases etc.). The graph of A_{Δ_1} cos α_{33} plotted by taking the higher scattering phases into account (3) is also given in Fig. 5 (dotted curve). As can be seen, in this case the calculated value of the energy E, for which $\cos\,\alpha_{33}$ passes through zero, is approximately 350 Mev, and this agrees with the value found from experiment.

Thus the experimental data on the photo-production of mesons can be explained quite simply and naturally if it is assumed that the main contribution to the cross-section comes from those matrix elements which lead to the formation of a meson – nucleon system with $I = J = 3/2$.

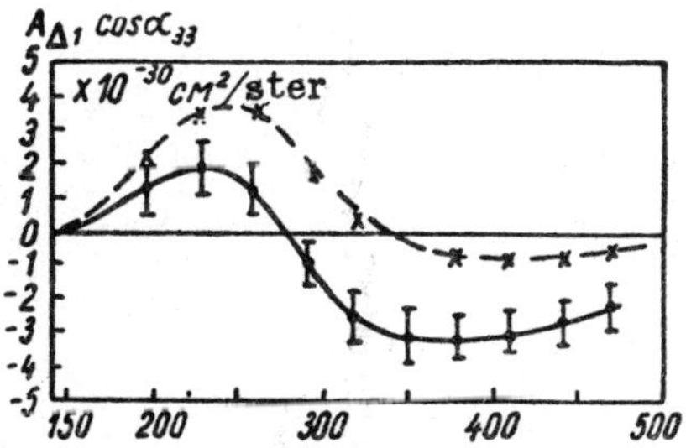

Fig. 5.

The values of E_d^+, M_{33}, E_{33} can be found with the help of the expressions $A = |E_d|^2$, (4.2a and 2b) and the data of Table 4. The relation between these quantities and the energy is depicted graphically in Fig. 6.

M_{33} and E_{33} are approximately equal to one another and a few times less than E_d^+ over the entire range of energy.

To determine the relative contribution of the γ-quanta of the different multipoles, let us consider the total cross-section of the process. For example, in the case of photo-production of π^+-mesons

$$\sigma^+ = 4\pi\left(A^+ + \frac{1}{3}C^+\right) = 4\pi\left[E_d^{+^2} + 4M_{33}^2 + \frac{1}{3}E_{33}^2 + \frac{2}{3}A_{\Delta_1}\cos\alpha_{33}\right] \tag{4.5a}$$

Since M_{33} and E_{33} are approximately equal over the entire range of energy (Fig. 6), it follows from the last expression that the contribution of M_{33} to the total cross-section σ^+ is about 12 times larger than that of E_{33}.

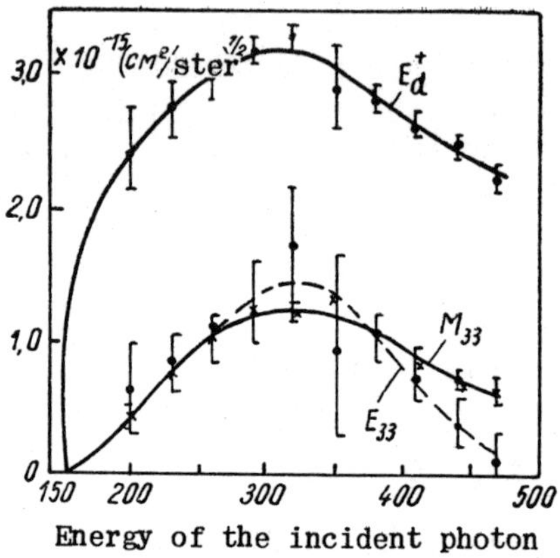

Fig. 6

2. *Energy dependence of the matrix elements* A_s, A_{k0}, A_p^+, A_{x0} This is given in Table 4. Let us try to describe this relation by means of an analytical expression, which must obviously depend on the momenta of the meson and photon. The "effective distance" z from the nucleon at which the meson is formed can also be introduced[1]. This means that we put $r = z$ in the radial wave function of the meson - nucleon system

$$\varphi = [j_1(\eta z)\cos\alpha_{33} - h_1(\eta z)\sin\alpha_{33}], \tag{4.5b}$$

where j_1 and h_1 are regular and irregular Bessel functions

η is the momentum of the meson in the CM system in units of μ c.

---

1 Of course z should not be literally taken as the distance at which the meson is formed. It is rather the most probable distance.

Instead of the function ϕ it is convenient to introduce

$$F(z) = \frac{\frac{\varphi}{\eta}}{\left(\frac{\varphi}{\eta}\right)_{\eta=0}}. \qquad (4.5c)$$

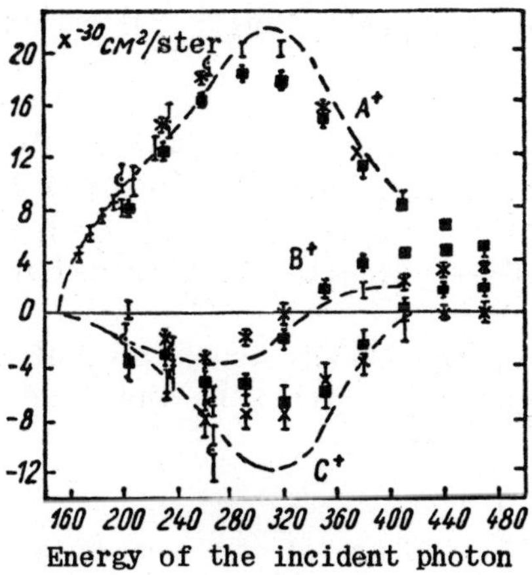

Fig. 7

We can then try to describe the dependence of the coefficients A_s etc. on the quantities mentioned by expressions of the following type:

$$\begin{aligned} A_s &= \rho g_s; \\ A_{x0} &= \nu^n \eta^2 \rho F^2 g_{x0}; \\ A_{k0} &= \nu^n \eta^2 \rho F^2 g_{k0}; \\ A_{\Delta_1} &= \eta^2 \rho F g_{\Delta_1}, \end{aligned} \qquad (4.6)$$

where ν is the momentum of the photon in units of μc in the CM system:

$$\rho \equiv \frac{\eta}{\nu}\left\{\left[1 + \nu\frac{\mu}{M}\right]\left[1 + \frac{\mu}{M}\sqrt{1+\eta^2}\right]\right\}^{-1}; \qquad (4.6a)$$

M, μ are the masses of the nucleon and the mèson;

$g_{x0}, g_{k0}, g_{\Delta_1}$ are numerical constants.

The unknown quantities in the expressions (4.6) are n, z and the constants g. In order to determine them, (4.6) has to be compared with the data of Table 4.

The values found for the coefficients g for the case $n = 0$ and $z = ½$ are given in Table 5. These values vary slightly when z varies from 0 to 1.

As an example the experimental values for A^+, B^+ and C^+ are compared in Fig. 7 with those calculated from (4.3) with the help of (4.6) for the case when $n = 0$, $z = ½$ and g are determined from Table 5. Similar results are obtained for A^0, C^0.

As can be seen, the energy dependence of the matrix elements can be described quite well with the help of the expressions (4.6).

TABLE 5
Values of the constants
g (10^{-30} cm^2/ster)

$g_s = 16{,}1 \pm 0{,}6;$	$g_{k0} = 0{,}41 \pm 0{,}11;$
$g_{x0} = 6{,}15 \pm 0{,}38;$	$g_{\Delta_1} = 1{,}42 \pm 0{,}37.$

3. *Polarization of the recoil neutrons during photo-production of mesons* (3). The correctness of the hypothesis of "resonance" interaction of the meson – nucleon system in the state $I = J = 3/2$ can also be verified by measuring the polarization of the recoil nucleons formed during the production of a meson.

The polarization of the recoil nucleons is given by

$$P = \frac{d\sigma_+ - d\sigma_-}{d\sigma_+ + d\sigma_-}, \qquad (4.7)$$

where $d\sigma_+$ and $d\sigma_-$ are the cross-sections of the photo-production of mesons, as a result of which a recoil nucleon is formed with its spin directed upwards and downwards respectively with respect to the production plane.

The expression for P is written as[1]

$$P = \frac{\sin\vartheta}{\sigma(\vartheta)}\left\{\operatorname{Im}\left[E_d^*\left(2M_d\,(^1/_2) + M_d\,(^3/_2) - \frac{1}{2}E_q\right)\right] + \right. \tag{4.8}$$

$$\left. + 3\operatorname{Im}\left[M_d^*\,(^1/_2)\left(M_d\,(^3/_2) - \frac{1}{2}E_q\right)\right]\cos\vartheta\right\},$$

where $\sigma(\theta)$ is the differential cross-section defined by the expression (1.3).

As can be seen, the expressions M_d (3./2) and E_q, the principal terms of which are M_{33} and E_{33} respectively, enter the expression (4.8) in the combination $M_d(3/2) - \frac{1}{2}E_q$. Since $M_{33} \approx E_{33}$ (Fig. 6), this difference is small and the polarization must be small.

Retaining only the resonance terms, we shall have, instead of (4.8),

$$P^+ = \frac{\sin\vartheta}{\sqrt{2}\sigma^+(\vartheta)}(A_s A_{k0})^{1/2}\left\{\frac{1}{3}[\sin(\alpha_{33} - \alpha_3) + 2\sin(\alpha_{33} - \alpha_1)]\right\} \tag{4.9}$$

$$P^0 = \frac{\sin\vartheta}{\sigma^0(\vartheta)}(A_s A_{k0})^{1/2}\left\{\frac{1}{3}[\sin(\alpha_{33} - \alpha_3) - \sin(\alpha_{33} - \alpha_1)]\right\}, \tag{4.10}$$

where A_s and A_{k0} are the same as in (4.2).

It is more convenient to make the measurements for $\nu \approx 90^\circ$, as in this interval of angles the contribution from the "non-resonance" phases is small.

The energy dependence of the polarization of the recoil nucleons for $\nu \approx 90^\circ$ is depicted in Fig. 8. As can be seen, for the case $\gamma + p \rightarrow \pi^+ + n$ this quantity increases with energy, but unfor-

1. The derivation of the expression for P (see (20)), with the help of which (4.8) can be obtained, is given in appendix 4; see also (21).

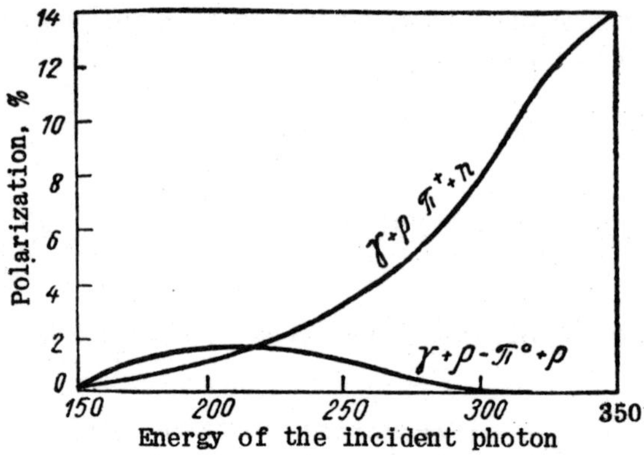

Fig. 8

tunately it remains quite small right up to $E\gamma \approx 350$ Mev. The polarization of recoil protons during the formation of neutral mesons ($\gamma + p \to p + \pi^0$) is even less ($<2\%$).

APPENDIX 1

TABLES OF CLEBSCH-GORDAN COEFFICIENTS

The wave function of the system $\psi_{c\gamma}$ with angular momentum C and its projection γ, consisting of two non-interacting subsystems (described by the functions $\psi_{a\alpha}$ and $\psi_{b\beta}$ with the angular momenta a and b and their projections α and β, can be written as:

$$\psi_{c\gamma} = \sum_{\gamma=\alpha+\beta} C^{c\gamma}_{a\alpha,\, b\beta} \psi_{a\alpha} \psi_{b\beta}. \qquad \text{(A.1)}$$

The coefficients $C^{c\gamma}_{a\alpha,\, b\beta}$ in this expression are called Clebsch-Gordan coefficients.[1] The subscripts indicate the angular momenta of the components (Roman letters) and their projections (Greek letters), the superscripts the total angular momentum and its projection. The Clebsch-Gordan coefficients characterize the relative importance of a state with given a and b and projections α and β.

The following is the explicit form of Clebsch-Gordan coefficients (12).

$$C^{c\gamma}_{a\alpha,\, b\beta} = \sqrt{\frac{(c+a-b)!\,(c-a+b)!\,(a+b-c)!\,(c+\gamma)!\,(c-\gamma)!\,(2c+1)}{(c+a+b+1)!\,(a-\alpha)!\,(a+\alpha)!\,(b-\beta)!\,(b+\beta)!}} \times$$

$$\times \sum_k \frac{(-1)^{k+b+\beta}\,(c+b+\alpha-k)!\,(a-\alpha+k)!\,\delta(\gamma, \alpha+\beta)}{(c-a+b-k)!\,(c+\gamma-k)!\,k!\,(k+a-b-\gamma)!}. \qquad \text{(A.2)}$$

Clebsch-Gordan coefficients possess the following properties:

---

1 Clebsch-Gordan coefficients as often denoted as: $(ab\alpha\beta|abc\gamma)$

a) orthogonality

$$\sum_{\alpha\beta} C^{c\gamma}_{a\alpha,\, b\beta} C^{c'\gamma'}_{a\alpha,\, b\beta} = \delta_{cc'}\,\delta_{\gamma\gamma'};$$

$$\sum_{c\alpha} C^{c\gamma}_{a\alpha,\, b\beta} C^{c\gamma}_{a\alpha,\, b'\beta'} = \frac{2c+1}{2b+1}\,\delta_{bb'}\delta_{\beta\beta'}; \qquad \text{(A. 3)}$$

b) symmetry

$$\begin{aligned} C^{c\gamma}_{a\alpha,\, b\beta} &= C^{c-\gamma}_{b-\beta,\, a-\alpha} = \\ &= (-1)^{a+b-c} C^{c\gamma}_{b\beta,\, a\alpha} = \\ &= (-1)^{a+b-c} C^{c-\gamma}_{a-\alpha,\, b-\beta} = \\ &= (-1)^{a-\alpha}\sqrt{\frac{2c+1}{2b+1}}\, C^{b-\beta}_{a\alpha,\, c-\gamma} = \\ &= (-1)^{b+\beta}\sqrt{\frac{2c+1}{2a+1}}\, C^{a-\alpha}_{c-\gamma,\, b\beta}. \end{aligned} \qquad \text{(A. 4)}$$

In the particular case when $\alpha = \beta = \gamma = 0$, we find from (A2)

$$C^{c0}_{a0,\, b0} = \frac{(-1)^{g-c}}{(g-a)!\,(g-b)!\,(g-c)!}\sqrt{\frac{(a+b-c)!\,(a-b+c)!\,(b+c-a)!\,(2c+1)}{(a+b+c+1)!}} =$$

$$= \frac{(-1)^{g-c}\, g!}{(g-a)!\,(g-b)!\,(g-c)!}\sqrt{\frac{2c+1}{(2g+1)!}(2g-2a)!\,(2g-2b)!\,(2g-2c)!}, \qquad \text{(A. 5)}$$

where

$$g = \frac{a+b+c}{2}. \qquad \text{(A. 5a)}$$

If $c = 0$, $a = b$, $\alpha = -\beta$, (A. 2) gives

$$C^{00}_{a\alpha,\, a-\beta} = (-1)^{a-\alpha}(2a+1)^{-1/2}. \qquad \text{(A. 5b)}$$

With the help of Clebsch-Gordan coefficients the product of two spherical harmonics can be expressed by means of a sum of single spherical harmonics:

$$Y^{*}_{lm}(\vartheta, \varphi)\, Y_{l'm'}(\vartheta, \varphi) = (-1)^{m} \sum_{n=|l-l'|}^{l+l'} \sqrt{\frac{(2l+1)(2l'+1)}{4\pi(2n+1)}} \times$$

$$\times\, C^{n0}_{l0,\, l'0} C^{nM}_{l-m,\, l'm'} Y_{nM}(\vartheta, \varphi). \tag{A.6}$$

The analytical expression for Clebsch-Gordan coefficients for b = ½, 1, 3/2 and 2 are given in the paper (26), for b = 5/2 in the paper (27), for b = 3 in the papers (28, 29).

There are tables of the numerical values of Clebsch-Gordan coefficients (30). Some of these values are given in Table 6[1]. In the use of this table it should be borne in mind that, if the required coefficient is not in the table, it may be transformed with the help of the relations (A.2) to one of the coefficients in the table.

1 In table (30) the values of Clebsch-Gordan coefficients for b from ½ to 9/2 inclusively are given to ten places. The values of the coefficients given by us contain only three figures.

TABLE 6

Numerical values of Clebsch-Gordan coefficients for b = ½, 1, 3/2, 2, 5/2, 3

β	γ	a	b	c	C
$-1/2$	0	$1/2$	$1/2$	1	0.707
$1/2$	0	$1/2$	$1/2$	1	0.707
$1/2$	1	$1/2$	$1/2$	1	1.000
-1	0	1	1	1	0.707
0	0	1	1	1	0.000
1	0	1	1	1	-0.707
0	1	1	1	1	0.707
1	1	1	1	1	-0.707
0	$1/2$	$1/2$	1	$3/2$	0.817
1	$1/2$	$1/2$	1	$3/2$	0.577
1	$3/2$	$1/2$	1	$3/2$	1.000
$-1/2$	$1/2$	1	$3/2$	$3/2$	0.730
$1/2$	$1/2$	1	$3/2$	$3/2$	-0.258
$3/2$	$1/2$	1	$3/2$	$3/2$	-0.633
$1/2$	$3/2$	1	$3/2$	$3/2$	0.633
$3/2$	$3/2$	1	$3/2$	$3/2$	-0.775
-1	0	1	1	2	0.408
0	0	1	1	2	0.817
1	0	1	1	2	0.408
0	1	1	1	2	0.707
1	1	1	1	2	0.707
1	2	1	1	2	1.000
$-1/2$	0	$1/2$	$3/2$	2	0.707
$1/2$	0	$1/2$	$3/2$	2	0.707
$1/2$	1	$1/2$	$3/2$	2	0.866
$3/2$	1	$1/2$	$3/2$	2	0.500
$3/2$	2	$1/2$	$3/2$	2	1.000
$-3/2$	0	$3/2$	$3/2$	2	0.500
$-1/2$	0	$3/2$	$3/2$	2	0.500
$1/2$	0	$3/2$	$3/2$	2	-0.500
$3/2$	0	$3/2$	$3/2$	2	-0.500
$-1/2$	1	$3/2$	$3/2$	2	0.707
$1/2$	1	$3/2$	$3/2$	2	0.000
$3/2$	1	$3/2$	$3/2$	2	-0.707
$1/2$	2	$3/2$	$3/2$	2	0.707
$3/2$	2	$3/2$	$3/2$	2	-0.707
-1	0	1	2	2	0.707
0	0	1	2	2	0.000
1	0	1	2	2	-0.707

Continuation of Table 6

β	γ	*a*	*b*	*c*	*C*
0	1	1	2	2	0.707
1	1	1	2	2	—0.408
2	1	1	2	2	—0.577
1	2	1	2	2	0.577
2	2	1	2	2	—0.817
—2	0	2	2	2	0.535
—1	0	2	2	2	0.267
0	0	2	2	2	—0.535
1	0	2	2	2	0.267
2	0	2	2	2	0.535
—1	1	2	2	2	0.655
0	1	2	2	2	—0.267
1	1	2	2	2	—0.267
2	1	2	2	2	0.655
0	2	2	2	2	0.535
1	2	2	2	2	—0.655
2	2	2	2	2	0.535
$-^1/_2$	$^1/_2$	1	$^3/_2$	$^5/_2$	0.548
$^1/_2$	$^1/_2$	1	$^3/_2$	$^5/_2$	0.775
$^3/_2$	$^1/_2$	1	$^3/_2$	$^5/_2$	0.316
$^1/_2$	$^3/_2$	1	$^3/_2$	$^5/_2$	0.775
$^3/_2$	$^3/_2$	1	$^3/_2$	$^5/_2$	0.632
$^3/_2$	$^5/_2$	1	$^3/_2$	$^5/_2$	1.000
0	$^1/_2$	$^1/_2$	2	$^5/_2$	0.775
1	$^1/_2$	$^1/_2$	2	$^5/_2$	0.632
1	$^3/_2$	$^1/_2$	2	$^5/_2$	0.894
2	$^3/_2$	$^1/_2$	2	$^5/_2$	0.447
2	$^5/_2$	$^1/_2$	2	$^5/_2$	1.000
—1	$^1/_2$	$^3/_2$	2	$^5/_2$	0.621
0	$^1/_2$	$^3/_2$	2	$^5/_2$	0.293
1	$^1/_2$	$^3/_2$	2	$^5/_2$	—0.598
2	$^1/_2$	$^3/_2$	2	$^5/_2$	—0.414
0	$^3/_2$	$^3/_2$	2	$^5/_2$	0.717
1	$^3/_2$	$^3/_2$	2	$^5/_2$	—0.169
2	$^3/_2$	$^3/_2$	2	$^5/_2$	—0.676
1	$^5/_2$	$^3/_2$	2	$^5/_2$	0.655
2	$^5/_2$	$^3/_2$	2	$^5/_2$	—0.756
$-^1/_2$	$^1/_2$	1	$^5/_2$	$^5/_2$	0.717
$^1/_2$	$^1/_2$	1	$^5/_2$	$^5/_2$	—0.169
$^3/_2$	$^1/_2$	1	$^5/_2$	$^5/_2$	—0.676
$^1/_2$	$^3/_2$	1	$^5/_2$	$^5/_2$	0.676

Continuation of Table 6

β	γ	a	b	c	C
3/2	3/2	1	5/2	5/2	—0.507
5/2	3/2	1	5/2	5/2	—0.535
3/2	5/2	1	5/2	5/2	0.535
5/2	5/2	1	5/2	5/2	—0.845
—3/2	1/2	2	5/2	5/2	0.621
—1/2	1/2	2	5/2	5/2	0.000
1/2	1/2	2	5/2	5/2	—0.478
3/2	1/2	2	5/2	5/2	0.414
5/2	1/2	2	5/2	5/2	0.463
—1/2	3/2	2	5/2	5/2	0.621
1/2	3/2	2	5/2	5/2	—0.414
3/2	3/2	2	5/2	5/2	—0.120
5/2	3/2	2	5/2	5/2	0.655
1/2	5/2	2	5/2	5/2	0.463
3/2	5/2	2	5/2	5/2	—0.655
5/2	5/2	2	5/2	5/2	0.598
—3/2	0	3/2	3/2	3	0.224
—1/2	0	3/2	3/2	3	0.671
1/2	0	3/2	3/2	3	0.671
3/2	0	3/2	3/2	3	0.224
—1/2	1	3/2	3/2	3	0.447
1/2	1	3/2	3/2	3	0.775
3/2	1	3/2	3/2	3	0.447
1/2	2	3/2	3/2	3	0.707
3/2	2	3/2	3/2	3	0.707
3/2	3	3/2	3/2	3	1.000
—1	0	1	2	3	0.447
0	0	1	2	3	0.775
1	0	1	2	3	0.447
0	1	1	2	3	0.633
1	1	1	2	3	0.730
2	1	1	2	3	0.258
1	2	1	2	3	0.817
2	2	1	2	3	0.577
2	3	1	2	3	1.000
—2	0	2	2	3	0.316
—1	0	2	2	3	0.633
0	0	2	2	3	0.000
1	0	2	2	3	—0.635
2	0	2	2	3	—0.316
—1	1	2	2	3	0.548

Continuation of Table 6

β	γ	a	b	c	C
0	1	2	2	3	0.447
1	1	2	2	3	—0.447
2	1	2	2	3	—0.548
0	2	2	2	3	0.707
1	2	2	2	3	0.000
2	2	2	2	3	—0.707
1	3	2	2	3	0.707
2	3	2	2	3	—0.707
$-1/2$	0	$1/2$	$5/2$	3	0.707
$1/2$	0	$1/2$	$5/2$	3	0.707
$1/2$	1	$1/2$	$5/2$	3	0.817
$3/2$	1	$1/2$	$5/2$	3	0.577
$3/2$	2	$1/2$	$5/2$	3	0.913
$5/2$	2	$1/2$	$5/2$	3	0.408
$5/2$	3	$1/2$	$5/2$	3	1.000
$-3/2$	0	$3/2$	$5/2$	3	0.548
$-1/2$	0	$3/2$	$5/2$	3	0.447
$1/2$	0	$3/2$	$5/2$	3	—0.447
$3/2$	0	$3/2$	$5/2$	3	—0.548
$-1/2$	1	$3/2$	$5/2$	3	0.671
$1/2$	1	$3/2$	$5/2$	3	0.129
$3/2$	1	$3/2$	$5/2$	3	—0.639
$5/2$	1	$3/2$	$5/2$	3	—0.354
$1/2$	2	$3/2$	$5/2$	3	0.707
$3/2$	2	$3/2$	$5/2$	3	—0.289
$5/2$	2	$3/2$	$5/2$	3	—0.645
$3/2$	3	$3/2$	$5/2$	3	0.612
$5/2$	3	$3/2$	$5/2$	3	—0.791
$-5/2$	0	$5/2$	$5/2$	3	0.373
$-3/2$	0	$5/2$	$5/2$	3	0.522
$-1/2$	0	$5/2$	$5/2$	3	—0.298
$1/2$	0	$5/2$	$5/2$	3	—0.298
$3/2$	0	$5/2$	$5/2$	3	0.522
$5/2$	0	$5/2$	$5/2$	3	0.373
$-3/2$	1	$5/2$	$5/2$	3	0.577
$-1/2$	1	$5/2$	$5/2$	3	0.183
$1/2$	1	$5/2$	$5/2$	3	—0.516
$3/2$	1	$5/2$	$5/2$	3	0.183
$5/2$	1	$5/2$	$5/2$	3	0.577
$-1/2$	2	$5/2$	$5/2$	3	0.646
$1/2$	2	$5/2$	$5/2$	3	—0.289

Continuation of Table 6

β	γ	a	b	c	C
$^3/_2$	2	$^5/_2$	$^5/_2$	3	—0.289
$^5/_2$	2	$^5/_2$	$^5/_2$	3	0.646
$^1/_2$	3	$^5/_2$	$^5/_2$	3	0.527
$^3/_2$	3	$^5/_2$	$^5/_2$	3	—0.667
$^5/_2$	3	$^5/_2$	$^5/_2$	3	0.527
—1	0	1	3	3	0.707
0	0	1	3	3	0.000
1	0	1	3	3	—0.707
0	1	1	3	3	0.707
1	1	1	3	3	—0.289
2	1	1	3	3	—0.646
1	2	1	3	3	0.646
2	2	1	3	3	—0.577
3	2	1	3	3	—0.500
2	3	1	3	3	0.500
3	3	1	3	3	—0.866
—2	0	2	3	3	0.577
—1	0	2	3	3	0.183
0	0	2	3	3	—0.516
1	0	2	3	3	0.183
2	0	2	3	3	0.577
—1	1	2	3	3	0.633
0	1	2	3	3	—0.183
1	1	2	3	3	—0.387
2	1	2	3	3	0.500
3	1	2	3	3	0.408
0	2	2	3	3	0.577
1	2	2	3	3	—0.500
2	2	2	3	3	0.000
3	2	2	3	3	0.645
1	3	2	3	3	0.408
2	3	2	3	3	—0.646
3	3	2	3	3	0.646
—3	0	3	3	3	0.408
—2	0	3	3	3	0.408
—1	0	3	3	3	—0.408
0	0	3	3	3	—0.000
1	0	3	3	3	0.408
2	0	3	3	3	—0.408
3	0	3	3	3	—0.408
—2	1	3	3	3	0.577

Continuation of Table 6

β	γ	a	b	c	C
—1	1	3	3	3	0.000
0	1	3	3	3	—0.408
1	1	3	3	3	0.408
2	1	3	3	3	0.000
3	1	3	3	3	—0.577
—1	2	3	3	3	0.577
0	2	3	3	3	—0.408
1	2	3	3	3	0.000
2	2	3	3	3	0.408
3	2	3	3	3	—0.577
0	3	3	3	3	0.408
1	3	3	3	3	—0.577
2	3	3	3	3	0.577
3	3	3	3	3	—0.408
—1	$^1/_2$	$^3/_2$	2	$^7/_2$	0.338
0	$^1/_2$	$^3/_2$	2	$^7/_2$	0.717
1	$^1/_2$	$^3/_2$	2	$^7/_2$	0.586
2	$^1/_2$	$^3/_2$	2	$^7/_2$	0.169
0	$^3/_2$	$^3/_2$	2	$^7/_2$	0.535
1	$^3/_2$	$^3/_2$	2	$^7/_2$	0.756
2	$^3/_2$	$^3/_2$	2	$^7/_2$	0.378
1	$^5/_2$	$^3/_2$	2	$^7/_2$	0.756
2	$^5/_2$	$^3/_2$	2	$^7/_2$	0.655
2	$^7/_2$	$^3/_2$	2	$^7/_2$	1.000
$-^1/_2$	$^1/_2$	1	$^5/_2$	$^7/_2$	0.535
$^1/_2$	$^1/_2$	1	$^5/_2$	$^7/_2$	0.756
$^3/_2$	$^1/_2$	1	$^5/_2$	$^7/_2$	0.378
$^1/_2$	$^3/_2$	1	$^5/_2$	$^7/_2$	0.690
$^3/_2$	$^3/_2$	1	$^5/_2$	$^7/_2$	0.690
$^5/_2$	$^3/_2$	1	$^5/_2$	$^7/_2$	0.218
$^3/_2$	$^5/_2$	1	$^5/_2$	$^7/_2$	0.845
$^5/_2$	$^5/_2$	1	$^5/_2$	$^7/_2$	0.535
$^5/_2$	$^7/_2$	1	$^5/_2$	$^7/_2$	1.000
$-^3/_2$	$^1/_2$	2	$^5/_2$	$^7/_2$	0.451
$-^1/_2$	$^1/_2$	2	$^5/_2$	$^7/_2$	0.558
$^1/_2$	$^1/_2$	2	$^5/_2$	$^7/_2$	—0.195
$^3/_2$	$^1/_2$	2	$^5/_2$	$^7/_2$	—0.620
$^5/_2$	$^1/_2$	2	$^5/_2$	$^7/_2$	—0.252
$-^1/_2$	$^3/_2$	2	$^5/_2$	$^7/_2$	0.617
$^1/_2$	$^3/_2$	2	$^5/_2$	$^7/_2$	0.309
$^3/_2$	$^3/_2$	2	$^5/_2$	$^7/_2$	—0.535

Continuation of Table 6

β	γ	a	b	c	C
5/2	3/2	2	5/2	7/2	—0.488
1/2	5/2	2	5/2	7/2	0.713
3/2	5/2	2	5/2	7/2	—0.126
5/2	5/2	2	5/2	7/2	—0.690
3/2	7/2	2	5/2	7/2	0.667
5/2	7/2	2	5/2	7/2	—0.745
0	1/2	1/2	3	7/2	0.756
1	1/2	1/2	3	7/2	0.655
1	3/2	1/2	3	7/2	0.845
2	3/2	1/2	3	7/2	0.535
2	5/2	1/2	3	7/2	0.926
3	5/2	1/2	3	7/2	0.378
3	7/2	1/2	3	7/2	1.000
—1	1/2	3/2	3	7/2	0.617
0	1/2	3/2	3	7/2	0.309
1	1/2	3/2	3	7/2	—0.535
2	1/2	3/2	3	7/2	—0.488
0	3/2	3/2	3	7/2	0.690
1	3/2	3/2	3	7/2	0.000
2	3/2	3/2	3	7/2	—0.655
3	3/2	3/2	3	7/2	0.309
1	5/2	3/2	3	7/2	0.690
2	5/2	3/2	3	7/2	—0.378
3	5/2	3/2	3	7/2	—0.617
2	7/2	3/2	3	7/2	0.577
3	7/2	3/2	3	7/2	—0.816
—2	1/2	5/2	2	7/2	0.504
—1	1/2	5/2	3	7/2	0.356
0	1/2	5/2	3	7/2	—0.436
1	1/2	5/2	3	7/2	—0.126
2	1/2	5/2	3	7/2	0.563
3	1/2	5/2	3	7/2	0.309
—1	3/2	5/2	3	7/2	0.617
0	3/2	5/2	3	7/2	0.000
1	3/2	5/2	3	7/2	—0.488
2	3/2	5/2	3	7/2	0.309
3	3/2	5/2	3	7/2	0.535
0	5/2	5/2	3	7/2	0.617
1	5/2	5/2	3	7/2	—0.398
2	5/2	5/2	3	7/2	—0.178
3	5/2	5/2	3	7/2	0.655

Continuation of Table 6

β	γ	*a*	*b*	*c*	*C*
1	$^7/_2$	$^5/_2$	3	$^7/_2$	0.471
2	$^7/_2$	$^5/_2$	3	$^7/_2$	—0.667
3	$^7/_2$	$^5/_2$	3	$^7/_2$	0.577
—2	0	2	2	4	0.120
—1	0	2	2	4	0.478
0	0	2	2	4	0.717
1	0	2	2	4	0.478
2	0	2	2	4	0.120
—1	1	2	2	4	0.267
0	1	2	2	4	0.655
1	1	2	2	4	0.655
2	1	2	2	4	0.267
0	2	2	2	4	0.463
1	2	2	2	4	0.756
2	2	2	2	4	0.463
1	3	2	2	4	0.707
2	3	2	2	4	0.707
2	4	2	2	4	1.000
—$^3/_2$	0	$^3/_2$	$^5/_2$	4	0.267
—$^1/_2$	0	$^3/_2$	$^5/_2$	4	0.655
$^1/_2$	0	$^3/_2$	$^5/_2$	4	0.655
$^3/_2$	0	$^3/_2$	$^5/_2$	4	0.267
—$^1/_2$	1	$^3/_2$	$^5/_2$	4	0.423
$^1/_2$	1	$^3/_2$	$^5/_2$	4	0.732
$^3/_2$	1	$^3/_2$	$^5/_2$	4	0.518
$^5/_2$	1	$^3/_2$	$^5/_2$	4	0.134
$^1/_2$	2	$^3/_2$	$^5/_2$	4	0.598
$^3/_2$	2	$^3/_2$	$^5/_2$	4	0.732
$^5/_2$	2	$^3/_2$	$^5/_2$	4	0.327
$^3/_2$	3	$^3/_2$	$^5/_2$	4	0.791
$^5/_2$	3	$^3/_2$	$^5/_2$	4	0.612
$^5/_2$	4	$^3/_2$	$^5/_2$	4	1.000
—$^5/_2$	0	$^5/_2$	$^5/_2$	4	0.189
—$^3/_2$	0	$^5/_2$	$^5/_2$	4	0.567
—$^1/_2$	0	$^5/_2$	$^5/_2$	4	0.378
$^1/_2$	0	$^5/_2$	$^5/_2$	4	—0.378
$^3/_2$	0	$^5/_2$	$^5/_2$	4	—0.567
$^5/_2$	0	$^5/_2$	$^5/_2$	4	—0.189
—$^3/_2$	1	$^5/_2$	$^5/_2$	4	0.378
—$^1/_2$	1	$^5/_2$	$^5/_2$	4	0.598
$^1/_2$	1	$^5/_2$	$^5/_2$	4	0.000

Continuation of Table 6

β	γ	a	b	c	C
$^3/_2$	1	$^5/_2$	$^5/_2$	4	—0.598
$^5/_2$	1	$^5/_2$	$^5/_2$	4	—0.378
—$^1/_2$	2	$^5/_2$	$^5/_2$	4	0.567
$^1/_2$	2	$^5/_2$	$^5/_2$	4	0.423
$^3/_2$	2	$^5/_2$	$^5/_2$	4	—0.423
$^5/_2$	2	$^5/_2$	$^5/_2$	4	—0.567
$^1/_2$	3	$^5/_2$	$^5/_2$	4	0.707
$^3/_2$	3	$^5/_2$	$^5/_2$	4	0.000
$^5/_2$	3	$^5/_2$	$^5/_2$	4	—0.707
$^3/_2$	4	$^5/_2$	$^5/_2$	4	0.707
$^5/_2$	4	$^5/_2$	$^5/_2$	4	—0.707
—1	0	1	3	4	0.463
0	0	1	3	4	0.756
1	0	1	3	4	0.463
0	1	1	3	4	0.598
1	1	1	3	4	0.732
2	1	1	3	4	0.327
1	2	1	3	4	0.732
2	2	1	3	4	0.655
3	2	1	3	4	0.189
2	3	1	3	4	0.866
3	3	1	3	4	0.500
3	4	1	3	4	1.000
—2	0	2	3	4	0.378
—1	0	2	3	4	0.598
0	0	2	3	4	0.000
1	0	2	3	4	—0.598
2	0	2	3	4	—0.378
—1	1	2	3	4	0.535
0	1	2	3	4	0.463
1	1	2	3	4	—0.327
2	1	2	3	4	—0.592
3	1	2	3	4	—0.207
0	2	2	3	4	0.655
1	2	2	3	4	0.189
2	2	2	3	4	—0.586
3	2	2	3	4	—0.439
1	3	2	3	4	0.707
2	3	2	3	4	—0.224
3	3	2	3	4	—0.671
2	4	2	3	4	0.633

Continuation of Table 6

β	γ	*a*	*b*	*c*	*C*
3	4	2	3	4	—0.775
—3	0	3	3	4	0.242
—2	0	3	3	4	0.564
—1	0	3	3	4	0.081
0	0	3	3	4	—0.484
1	0	3	3	4	0.081
2	0	3	3	4	0.564
3	0	3	3	4	0.242
—2	1	3	3	4	0.441
—1	1	3	3	4	0.456
0	1	3	3	4	—0.312
1	1	3	3	4	—0.312
2	1	3	3	4	0.456
3	1	3	3	4	0.441
—1	2	3	3	4	0.592
0	2	3	3	4	0.140
1	2	3	3	4	—0.510
2	2	3	3	4	0.140
3	2	3	3	4	0.592
0	3	3	3	4	0.640
1	3	3	3	4	—0.302
2	3	3	3	4	—0.302
3	3	3	3	4	0.640
1	4	3	3	4	0.522
2	4	3	3	4	—0.674
3	4	3	3	4	0.522
—$^3/_2$	$^1/_2$	2	$^5/_2$	$^9/_2$	0.199
—$^1/_2$	$^1/_2$	2	$^5/_2$	$^9/_2$	0.563
$^1/_2$	$^1/_2$	2	$^5/_2$	$^9/_2$	0.690
$^3/_2$	$^1/_2$	2	$^5/_2$	$^9/_2$	0.398
$^5/_2$	$^1/_2$	2	$^5/_2$	$^9/_2$	0.089
—$^1/_2$	$^3/_2$	2	$^5/_2$	$^9/_2$	0.345
$^1/_2$	$^3/_2$	2	$^5/_2$	$^9/_2$	0.690
$^3/_2$	$^3/_2$	2	$^5/_2$	$^9/_2$	0.598
$^5/_2$	$^3/_2$	2	$^5/_2$	$^6/_2$	0.218
$^1/_2$	$^5/_2$	2	$^5/_2$	$^9/_2$	0.527
$^3/_2$	$^5/_2$	2	$^5/_2$	$^9/_2$	0.745
$^5/_2$	$^5/_2$	2	$^5/_2$	$^9/_2$	0.408
$^3/_2$	$^7/_2$	2	$^5/_2$	$^9/_2$	0.745
$^5/_2$	$^7/_2$	2	$^5/_2$	$^9/_2$	0.667
$^5/_2$	$^9/_2$	2	$^5/_2$	$^9/_2$	1.000

Continuation of Table 6

β	γ	*a*	*b*	*c*	*C*
—1	$^1/_2$	$^3/_2$	3	$^9/_2$	0.345
0	$^1/_2$	$^3/_2$	3	$^9/_2$	0.690
1	$^1/_2$	$^3/_2$	3	$^9/_2$	0.598
2	$^1/_2$	$^3/_2$	3	$^9/_2$	0.218
0	$^3/_2$	$^3/_2$	3	$^9/_2$	0.488
1	$^3/_2$	$^3/_2$	3	$^9/_2$	0.732
2	$^3/_2$	$^3/_2$	3	$^9/_2$	0.463
3	$^3/_2$	$^3/_2$	3	$^9/_2$	0.109
1	$^5/_2$	$^3/_2$	3	$^9/_2$	0.646
2	$^5/_2$	$^3/_2$	3	$^9/_2$	0.707
3	$^5/_2$	$^3/_2$	3	$^9/_2$	0.289
2	$^7/_2$	$^3/_2$	3	$^9/_2$	0.816
3	$^7/_2$	$^3/_2$	3	$^9/_2$	0.577
3	$^9/_2$	$^3/_2$	3	$^9/_2$	1.000
—2	$^1/_2$	$^5/_2$	3	$^9/_2$	0.300
—1	$^1/_2$	$^5/_2$	3	$^9/_2$	0.595
0	$^1/_2$	$^5/_2$	3	$^9/_2$	0.208
1	$^1/_2$	$^5/_2$	3	$^9/_2$	—0.481
2	$^1/_2$	$^5/_2$	3	$^9/_2$	—0.510
3	$^1/_2$	$^5/_2$	3	$^9/_2$	—0.147
—1	$^3/_2$	$^5/_2$	3	$^9/_2$	0.465
0	$^3/_2$	$^5/_2$	3	$^9/_2$	0.541
1	$^3/_2$	$^5/_2$	3	$^9/_2$	—0.147
2	$^3/_2$	$^5/_2$	3	$^9/_2$	—0.605
3	$^3/_2$	$^5/_2$	3	$^9/_2$	—0.322
0	$^5/_2$	$^5/_2$	3	$^9/_2$	0.616
1	$^5/_2$	$^5/_2$	3	$^9/_2$	0.318
2	$^5/_2$	$^5/_2$	3	$^9/_2$	—0.498
3	$^5/_2$	$^5/_2$	3	$^9/_2$	—0.522
1	$^7/_2$	$^5/_2$	3	$^9/_2$	0.711
2	$^7/_2$	$^5/_2$	3	$^9/_2$	—0.101
3	$^7/_2$	$^5/_2$	3	$^9/_2$	—0.696
2	$^9/_2$	$^5/_2$	3	$^9/_2$	0.674
3	$^9/_2$	$^5/_2$	3	$^9/_2$	—0.739

TABLES OF RACAH COEFFICIENTS AND Z-COEFFICIENTS

Racah coefficients appear when the wave-function is formed for a system consisting of three non-interacting sub-systems (22).

The explicit form of Racah coefficients is as follows:

$$W(abcd;\ ef) = \Delta(abe)\,\Delta(cde)\,\Delta(acf)\,\Delta(bdf)\,w(abcd;\ ef), \qquad \text{(A.7a)}$$

where

$$w(abcd;\ ef) = \sum_z \frac{(-1)^{z+a+b+c+d}\,(z+1)!}{(z-a-b-e)!\,(z-c-d-e)!\,(z-a-c-f)!\,(z-b-d-f)!} \times$$

$$\times \frac{1}{(a+b+c+d-z)!\,(a+d+e+f-z)!\,(b+c+e+f-z)!}; \qquad \text{(A.7b)}$$

$$\Delta(abc) = \left[\frac{(a+b-c)!\,(a-b+c)!\,(b+c-a)!}{(a+b+c+1)!}\right]^{1/2}.$$

Racah coefficients possess the following symmetry properties:

$$W(abcd;\ ef) = W(badc;\ ef) = W(cdab;\ ef) = W(acbd;\ fe) =$$
$$= (-1)^{e+f-a-d}\,W(ebcf;\ ad) = (-1)^{e+f-b-c}\,W(aefd;\ bc). \qquad \text{(A.8)}$$

Moreover, every numerical triplet (abe), (cde), (acf), (bdf) must satisfy the following condition: any of these three numbers must be less than or equal to the sum of the two others. Otherwise the Racah coefficient becomes zero.

In the particular case $e = 0$

$$W(abcd;\ 0f) = (-1)^{b+c-f} \frac{\delta_{ab}\delta_{cd}}{\sqrt{(2b+1)(2c+1)}}. \tag{A.9}$$

If $f = 0$

$$W(abcd;\ e0) = (-1)^{c+b-e} \frac{\delta_{ac}\delta_{bd}}{\sqrt{(2b+1)(2c+1)}}. \tag{A.10}$$

The sum of the product of three Clebsch-Gordan coefficients over any projection, for example β, can be expressed in terms of Racah coefficients:

$$\sum_{\beta} C^{c\gamma}_{a\alpha,\ b\beta} C^{e\varepsilon}_{d\delta b\beta} C^{d\delta}_{a\alpha f\varphi} = \sqrt{(2c+1)(2d+1)}\, C^{e\varepsilon}_{c\gamma,\ f\varphi} W(cbfd;\ ae). \tag{A.11}$$

It is characteristic of Racah coefficients that they depend only on the angular momenta and not on their projections. For the reader's information we give certain relationships between Racah coefficients.

$$\sum_{e} (2e+1)(2f+1) W(abcd;\ ef) W(abcd;\ eg) = \delta_{fg};$$

$$W(agfb;\ cd) = \sum_{e} (2e+1)(-1)^{a+b-e} W(abcd;\ ef) W(bacd;\ eg);$$

$$W(a\alpha b\beta;\ c\gamma) W(a'\alpha b'\beta;\ c'\gamma) =$$

$$= \sum_{\lambda} (2\lambda+1) W(a'\lambda\alpha c;\ ac') W(b\lambda\beta c';\ b'c) W(a'\lambda\gamma b;\ ab'); \tag{A.12}$$

$$C^{e\alpha+\beta}_{a\alpha b\beta} C^{c\alpha+\beta+\delta}_{e\alpha+\beta d\delta} = \sum_{f} (2e+1)^{1/2} (2f+1)^{1/2} C^{f\beta+\delta}_{b\beta d\delta} C^{c\alpha+\beta+\delta}_{a\alpha f\beta+\delta} W(abcd;\ ef);$$

$$W(a\alpha + {}^1/_2;\ b\beta + {}^1/_2;\ c + {}^1/_2\gamma) =$$

$$= (2c+1) \sqrt{\frac{(\alpha+\beta+\gamma+2)(\alpha+\beta-\gamma+1)}{(\alpha+c+a+2)(\alpha+c+1-a)(\beta+b+c+2)(\beta+c+1-b)}} \times$$

$$\times W(a\alpha b\beta;\ c\gamma) +$$

$$+ \sqrt{\frac{(a+\alpha+1-c)(a+c-\alpha)(b+\beta+1-c)(b+c-\beta)}{(\alpha+a+c+2)(\alpha+c+1-a)(\beta+b+c+2)(\beta+c+1-b)}} \times$$

$$\times W(a\alpha + {}^1/_2 b\beta + {}^1/_2;\ c,\ -{}^1/_2\gamma). \tag{A.12a}$$

It should also be noted that

$$Z(abcd;\ 0f) = \delta_{ab}\delta_{cd}(-1)^{2f}\, i^{f+c-a}\sqrt{(2a+1)(2c+1)}\,C^{f0}_{a0c0} \quad \text{(A. 12b)}$$

and

$$Z(abcd;\ e0) = \delta_{ac}\delta_{bd}(-1)^{b-e}\sqrt{2b+1}. \quad \text{(A. 12c)}$$

Finally,

$$Z(l_1J_1l_2J_2;\ sn) = (-1)^n Z(l_2J_2l_1J_1;\ sn) \quad \text{(A. 12d)}$$

and

$$\sum_b Z(abcd;\ ef)\,Z(abc'd;\ ef) = \delta_{cc'}(2a+1)(2d+1)\left[C^{f0}_{a0c0}\right]^2. \quad \text{(A. 12e)}$$

Tables of numerical values of Racah coefficients (24) and Z-coefficients (25) are in existence. Some of these values are given in Tables 7-10. The values of Racah coefficients for e from 0 to 3 and f from 0 to 8 inclusive are given up to ten places in the tables (24). We have confined ourselves to three figures only. The values of Z^2 $(l_1J_1l_2J_2;\ ef)$ for e from ½ to 3 and f from 0 to 8 inclusive are given in the tables (25).

TABLE 7

Numerical values of Racah coefficients $W(abcd;\ \tfrac{1}{2}f)$ for $f=0,1,2,3,4,5$

a	b	c	d	e	f	W
$^1/_2$	0	$^1/_2$	0	$^1/_2$	0	0.707
0	$^1/_2$	0	$^1/_2$	$^1/_2$	0	0.707
1	$^1/_2$	1	$^1/_2$	$^1/_2$	0	—0.408
$^1/_2$	1	$^1/_2$	1	$^1/_2$	0	—0.408
$^3/_2$	1	$^3/_2$	1	$^1/_2$	0	0.289
1	$^3/_2$	1	$^3/_2$	$^1/_2$	0	0.289
2	$^3/_2$	2	$^3/_2$	$^1/_2$	0	—0.224
$^3/_2$	2	$^3/_2$	2	$^1/_2$	0	—0.224
$^5/_2$	2	$^5/_2$	2	$^1/_2$	0	0.183
2	$^5/_2$	2	$^5/_2$	$^1/_2$	0	0.183
3	$^5/_2$	3	$^5/_2$	$^1/_2$	0	—0.154
$^5/_2$	3	$^5/_2$	3	$^1/_2$	0	—0.154
$^7/_2$	3	$^7/_2$	3	$^1/_2$	0	0.134
3	$^7/_2$	3	$^7/_2$	$^1/_2$	0	0.134
4	$^7/_2$	4	$^7/_2$	$^1/_2$	0	—0.118
$^7/_2$	4	$^7/_2$	4	$^1/_2$	0	—0.118
$^9/_2$	4	$^9/_2$	4	$^1/_2$	0	0.105
4	$^9/_2$	4	$^9/_2$	$^1/_2$	0	0.105
5	$^9/_2$	5	$^9/_2$	$^1/_2$	0	—0.095
$^1/_2$	1	$^1/_2$	0	$^1/_2$	1	0.408
$^3/_2$	1	$^1/_2$	0	$^1/_2$	1	0.408
1	$^1/_2$	0	$^1/_2$	$^1/_2$	1	0.408
1	$^3/_2$	0	$^1/_2$	$^1/_2$	1	0.408
0	$^1/_2$	1	$^1/_2$	$^1/_2$	1	0.408
1	$^1/_2$	1	$^1/_2$	$^1/_2$	1	0.333
1	$^3/_2$	1	$^1/_2$	$^1/_2$	1	—0.167
2	$^3/_2$	1	$^1/_2$	$^1/_2$	1	—0.289
$^1/_2$	0	$^1/_2$	1	$^1/_2$	1	0.408
$^1/_2$	1	$^1/_2$	1	$^1/_2$	1	0.333
$^3/_2$	1	$^1/_2$	1	$^1/_2$	1	—0.167
$^3/_2$	2	$^1/_2$	1	$^1/_2$	1	—0.289
$^1/_2$	0	$^3/_2$	1	$^1/_2$	1	0.408
$^1/_2$	1	$^3/_2$	1	$^1/_2$	1	—0.167
$^3/_2$	1	$^3/_2$	1	$^1/_2$	1	—0.264
$^3/_2$	2	$^3/_2$	1	$^1/_2$	1	0.091
$^5/_2$	2	$^3/_2$	1	$^1/_2$	1	0.224
0	$^1/_2$	1	$^3/_2$	$^1/_2$	1	0.408
1	$^1/_2$	1	$^3/_2$	$^1/_2$	1	—0.167
1	$^3/_2$	1	$^3/_2$	$^1/_2$	1	—0.264
2	$^3/_2$	1	$^3/_2$	$^1/_2$	1	0.091
2	$^5/_2$	1	$^3/_2$	$^1/_2$	1	0.224
1	$^1/_2$	2	$^3/_2$	$^1/_2$	1	—0.289

Continuation of Table 7

a	*b*	*c*	*d*	*e*	*f*	*W*
1	$^3/_2$	2	$^3/_2$	$^1/_2$	1	0.091
2	$^3/_2$	2	$^3/_2$	$^1/_2$	1	0.212
2	$^5/_2$	2	$^3/_2$	$^1/_2$	1	—0.058
3	$^5/_2$	2	$^3/_2$	$^1/_2$	1	—0.183
$^1/_2$	1	$^3/_2$	2	$^1/_2$	1	—0.289
$^3/_2$	1	$^3/_2$	2	$^1/_2$	1	0.091
$^3/_2$	2	$^3/_2$	2	$^1/_2$	1	0.212
$^5/_2$	2	$^3/_2$	2	$^1/_2$	1	—0.058
$^5/_2$	3	$^3/_2$	2	$^1/_2$	1	—0.183
$^3/_2$	1	$^5/_2$	2	$^1/_2$	1	0.224
$^3/_2$	2	$^5/_2$	2	$^1/_2$	1	—0.058
$^5/_2$	2	$^5/_2$	2	$^1/_2$	1	—0.176
$^5/_2$	3	$^5/_2$	2	$^1/_2$	1	0.040
$^7/_2$	3	$^5/_2$	2	$^1/_2$	1	0.154
1	$^3/_2$	2	$^5/_2$	$^1/_2$	1	0.224
2	$^3/_2$	2	$^5/_2$	$^1/_2$	1	—0.058
2	$^5/_2$	2	$^5/_2$	$^1/_2$	1	—0.176
3	$^5/_2$	2	$^5/_2$	$^1/_2$	1	0.039
3	$^7/_2$	2	$^5/_2$	$^1/_2$	1	0.154
2	$^3/_2$	3	$^5/_2$	$^1/_2$	1	—0.183
2	$^5/_2$	3	$^5/_2$	$^1/_2$	1	0.040
3	$^5/_2$	3	$^5/_2$	$^1/_2$	1	0.151
3	$^7/_2$	3	$^5/_2$	$^1/_2$	1	—0.029
4	$^7/_2$	3	$^5/_2$	$^1/_2$	1	—0.134
$^3/_2$	2	$^5/_2$	3	$^1/_2$	1	—0.183
$^5/_2$	2	$^5/_2$	3	$^1/_2$	1	0.040
$^5/_2$	3	$^5/_2$	3	$^1/_2$	1	0.151
$^7/_2$	3	$^5/_2$	3	$^1/_2$	1	—0.029
$^7/_2$	4	$^5/_2$	3	$^1/_2$	1	—0.134
$^5/_2$	2	$^7/_2$	3	$^1/_2$	1	0.154
$^5/_2$	3	$^7/_2$	3	$^1/_2$	1	—0.029
$^7/_2$	3	$^7/_2$	3	$^1/_2$	1	—0.131
$^7/_2$	4	$^7/_2$	3	$^1/_2$	1	0.022
$^9/_2$	4	$^7/_2$	3	$^1/_2$	1	0.118
2	$^5/_2$	3	$^7/_2$	$^1/_2$	1	0.154
3	$^5/_2$	3	$^7/_2$	$^1/_2$	1	—0.029
3	$^7/_2$	3	$^7/_2$	$^1/_2$	1	—0.131
4	$^7/_2$	3	$^7/_2$	$^1/_2$	1	0.022
4	$^9/_2$	3	$^7/_2$	$^1/_2$	1	0.118
3	$^5/_2$	4	$^7/_2$	$^1/_2$	1	—0.134
3	$^7/_2$	4	$^7/_2$	$^1/_2$	1	0.022

Continuation of Table 7

a	b	c	d	e	f	W
4	$^7/_2$	4	$^7/_2$	$^1/_2$	1	0.116
4	$^9/_2$	4	$^7/_2$	$^1/_2$	1	—0.018
5	$^9/_2$	4	$^7/_2$	$^1/_2$	1	—0.105
$^5/_2$	3	$^7/_2$	4	$^1/_2$	1	—0.134
$^7/_2$	3	$^7/_2$	4	$^1/_2$	1	0.022
$^7/_2$	4	$^7/_2$	4	$^1/_2$	1	0.116
$^9/_2$	4	$^7/_2$	4	$^1/_2$	1	—0.018
$^7/_2$	3	$^9/_2$	4	$^1/_2$	1	0.118
$^7/_2$	4	$^9/_2$	4	$^1/_2$	1	—0.018
$^9/_2$	4	$^9/_2$	4	$^1/_2$	1	—0.104
3	$^7/_2$	4	$^9/_2$	$^1/_2$	1	0.118
4	$^7/_2$	4	$^9/_2$	$^1/_2$	1	—0.018
4	$^9/_2$	4	$^9/_2$	$^1/_2$	1	—0.104
5	$^9/_2$	4	$^9/_2$	$^1/_2$	1	0.014
4	$^7/_2$	5	$^9/_2$	$^1/_2$	1	—0.105
4	$^9/_2$	5	$^9/_2$	$^1/_2$	1	0.014
5	$^9/_2$	5	$^9/_2$	$^1/_2$	1	0.095
$^3/_2$	2	$^1/_2$	0	$^1/_2$	2	0.316
$^5/_2$	2	$^1/_2$	0	$^1/_2$	2	0.316
2	$^3/_2$	0	$^1/_2$	$^1/_2$	2	0.316
2	$^5/_2$	0	$^1/_2$	$^1/_2$	2	0.316
1	$^3/_2$	1	$^1/_2$	$^1/_2$	2	0.287
2	$^3/_2$	1	$^1/_2$	$^1/_2$	2	0.224
2	$^5/_2$	1	$^1/_2$	$^1/_2$	2	—0.150
3	$^5/_2$	1	$^1/_2$	$^1/_2$	2	—0.236
$^3/_2$	1	$^1/_2$	1	$^1/_2$	2	0.288
$^3/_2$	2	$^1/_2$	1	$^1/_2$	2	0.224
$^5/_2$	2	$^1/_2$	1	$^1/_2$	2	—0.149
$^5/_2$	3	$^1/_2$	1	$^1/_2$	2	—0.236
$^1/_2$	1	$^3/_2$	1	$^1/_2$	2	0.289
$^3/_2$	1	$^3/_2$	1	$^1/_2$	2	0.204
$^3/_2$	2	$^3/_2$	1	$^1/_2$	2	—0.158
$^5/_2$	2	$^3/_2$	1	$^1/_2$	2	—0.197
$^5/_2$	3	$^3/_2$	1	$^1/_2$	2	0.089
$^7/_2$	3	$^3/_2$	1	$^1/_2$	2	0.189
1	$^1/_2$	1	$^3/_2$	$^1/_2$	2	0.289
1	$^3/_2$	1	$^3/_2$	$^1/_2$	2	0.204
2	$^3/_2$	1	$^3/_2$	$^1/_2$	2	—0.158
2	$^5/_2$	1	$^3/_2$	$^1/_2$	2	—0.197
3	$^5/_2$	1	$^3/_2$	$^1/_2$	2	0.089
3	$^7/_2$	1	$^3/_2$	$^1/_2$	2	0.189

Continuation of Table 7

a	b	c	d	e	f	W
0	$^1/_2$	2	$^3/_2$	$^1/_2$	2	0.316
1	$^1/_2$	2	$^3/_2$	$^1/_2$	2	0.224
1	$^3/_2$	2	$^3/_2$	$^1/_2$	2	—0.158
2	$^3/_2$	2	$^3/_2$	$^1/_2$	2	—0.187
2	$^5/_2$	2	$^3/_2$	$^1/_2$	2	0.100
3	$^5/_2$	2	$^3/_2$	$^1/_2$	2	0.169
3	$^7/_2$	2	$^3/_2$	$^1/_2$	2	—0.060
4	$^7/_2$	2	$^3/_2$	$^1/_2$	2	—0.158
$^1/_2$	0	$^3/_2$	2	$^1/_2$	2	0.316
$^1/_2$	1	$^3/_2$	2	$^1/_2$	2	0.224
$^3/_2$	1	$^3/_2$	2	$^1/_2$	2	—0.158
$^3/_2$	2	$^3/_2$	2	$^1/_2$	2	—0.187
$^5/_2$	2	$^3/_2$	2	$^1/_2$	2	0.100
$^5/_2$	3	$^3/_2$	2	$^1/_2$	2	0.169
$^7/_2$	3	$^3/_2$	2	$^1/_2$	2	—0.060
$^7/_2$	4	$^3/_2$	2	$^1/_2$	2	—0.158
$^1/_2$	0	$^5/_2$	2	$^1/_2$	2	0.316
$^1/_2$	1	$^5/_2$	2	$^1/_2$	2	—0.149
$^3/_2$	1	$^5/_2$	2	$^1/_2$	2	—0.197
$^3/_2$	2	$^5/_2$	2	$^1/_2$	2	0.100
$^5/_2$	2	$^5/_2$	2	$^1/_2$	2	0.163
$^5/_2$	3	$^5/_2$	2	$^1/_2$	2	—0.069
$^7/_2$	3	$^5/_2$	2	$^1/_2$	2	—0.146
$^7/_2$	4	$^5/_2$	2	$^1/_2$	2	0.043
$^9/_2$	4	$^5/_2$	2	$^1/_2$	2	0.136
0	$^1/_2$	2	$^5/_2$	$^1/_2$	2	0.316
1	$^1/_2$	2	$^5/_2$	$^1/_2$	2	—0.149
1	$^3/_2$	2	$^5/_2$	$^1/_2$	2	—0.197
2	$^3/_2$	2	$^5/_2$	$^1/_2$	2	0.100
2	$^5/_2$	2	$^5/_2$	$^1/_2$	2	0.163
3	$^5/_2$	2	$^5/_2$	$^1/_2$	2	—0.069
3	$^7/_2$	2	$^5/_2$	$^1/_2$	2	—0.146
4	$^7/_2$	2	$^5/_2$	$^1/_2$	2	0.043
4	$^9/_2$	2	$^5/_2$	$^1/_2$	2	0.136
1	$^1/_2$	3	$^5/_2$	$^1/_2$	2	—0.236
1	$^3/_2$	3	$^5/_2$	$^1/_2$	2	0.089
2	$^3/_2$	3	$^5/_2$	$^1/_2$	2	0.169
2	$^5/_2$	3	$^5/_2$	$^1/_2$	2	—0.069
3	$^5/_2$	3	$^5/_2$	$^1/_2$	2	—0.143
3	$^7/_2$	3	$^5/_2$	$^1/_2$	2	0.051
4	$^7/_2$	3	$^5/_2$	$^1/_2$	2	0.129

Continuation of Table 7

a	b	c	d	e	f	W
4	$^9/_2$	3	$^5/_2$	$^1/_2$	2	—0.033
5	$^9/_2$	3	$^5/_2$	$^1/_2$	2	—0.120
$^1/_2$	1	$^5/_2$	3	$^1/_2$	2	—0.236
$^3/_2$	1	$^5/_2$	3	$^1/_2$	2	0.089
$^3/_2$	2	$^5/_2$	3	$^1/_2$	2	0.169
$^5/_2$	2	$^5/_2$	3	$^1/_2$	2	—0.069
$^5/_2$	3	$^5/_2$	3	$^1/_2$	2	—0.143
$^7/_2$	3	$^5/_2$	3	$^1/_2$	2	0.051
$^7/_2$	4	$^5/_2$	3	$^1/_2$	2	0.129
$^9/_2$	4	$^5/_2$	3	$^1/_2$	2	—0.033
$^3/_2$	1	$^7/_2$	3	$^1/_2$	2	0.189
$^3/_2$	2	$^7/_2$	3	$^1/_2$	2	—0.060
$^5/_2$	2	$^7/_2$	3	$^1/_2$	2	—0.146
$^5/_2$	3	$^7/_2$	3	$^1/_2$	2	0.051
$^7/_2$	3	$^7/_2$	3	$^1/_2$	2	0.126
$^7/_2$	4	$^7/_2$	3	$^1/_2$	2	—0.039
$^9/_2$	4	$^7/_2$	3	$^1/_2$	2	—0.114
1	$^3/_2$	3	$^7/_2$	$^1/_2$	2	0.189
2	$^3/_2$	3	$^7/_2$	$^1/_2$	2	—0.060
2	$^5/_2$	3	$^7/_2$	$^1/_2$	2	—0.146
3	$^5/_2$	3	$^7/_2$	$^1/_2$	2	0.051
3	$^7/_2$	3	$^7/_2$	$^1/_2$	2	0.126
4	$^7/_2$	3	$^7/_2$	$^1/_2$	2	—0.039
4	$^9/_2$	3	$^7/_2$	$^1/_2$	2	—0.114
5	$^9/_2$	3	$^7/_2$	$^1/_2$	2	0.026
2	$^3/_2$	4	$^7/_2$	$^1/_2$	2	—0.158
2	$^5/_2$	4	$^7/_2$	$^1/_2$	2	0.043
3	$^5/_2$	4	$^7/_2$	$^1/_2$	2	0.129
3	$^7/_2$	4	$^7/_2$	$^1/_2$	2	—0.039
4	$^7/_2$	4	$^7/_2$	$^1/_2$	2	—0.113
4	$^9/_2$	4	$^7/_2$	$^1/_2$	2	0.030
5	$^9/_2$	4	$^7/_2$	$^1/_2$	2	0.103
$^3/_2$	2	$^7/_2$	4	$^1/_2$	2	—0.158
$^5/_2$	2	$^7/_2$	4	$^1/_2$	2	0.043
$^5/_2$	3	$^7/_2$	4	$^1/_2$	2	0.129
$^7/_2$	3	$^7/_2$	4	$^1/_2$	2	—0.039
$^7/_2$	4	$^7/_2$	4	$^1/_2$	2	—0.113
$^9/_2$	4	$^7/_2$	4	$^1/_2$	2	0.030
$^5/_2$	2	$^9/_2$	4	$^1/_2$	2	0.136
$^5/_2$	3	$^9/_2$	4	$^1/_2$	2	—0.033
$^7/_2$	3	$^9/_2$	4	$^1/_2$	2	—0.114

Continuation of Table 7

a	b	c	d	e	f	W
$^7/_2$	4	$^9/_2$	4	$^1/_2$	2	0.030
$^9/_2$	4	$^9/_2$	4	$^1/_2$	2	0.102
2	$^5/_2$	4	$^9/_2$	$^1/_2$	2	0.136
3	$^5/_2$	4	$^9/_2$	$^1/_2$	2	—0.033
3	$^7/_2$	4	$^9/_2$	$^1/_2$	2	—0.114
4	$^7/_2$	4	$^9/_2$	$^1/_2$	2	0.030
4	$^9/_2$	4	$^9/_2$	$^1/_2$	2	0.102
5	$^9/_2$	4	$^9/_2$	$^1/_2$	2	—0.025
3	$^5/_2$	5	$^9/_2$	$^1/_2$	2	—0.120
3	$^7/_2$	5	$^9/_2$	$^1/_2$	2	0.026
4	$^7/_2$	5	$^9/_2$	$^1/_2$	2	0.103
4	$^9/_2$	5	$^9/_2$	$^1/_2$	2	—0.025
5	$^9/_2$	5	$^9/_2$	$^1/_2$	2	—0.093
$^5/_2$	3	$^1/_2$	0	$^1/_2$	3	0.267
$^7/_2$	3	$^1/_2$	0	$^1/_2$	3	0.267
3	$^5/_2$	0	$^1/_2$	$^1/_2$	3	0.267
3	$^7/_2$	0	$^1/_2$	$^1/_2$	3	0.267
2	$^5/_2$	1	$^1/_2$	$^1/_2$	3	0.236
3	$^5/_2$	1	$^1/_2$	$^1/_2$	3	0.178
3	$^7/_2$	1	$^1/_2$	$^1/_2$	3	—0.134
4	$^7/_2$	1	$^1/_2$	$^1/_2$	3	—0.204
$^5/_2$	2	$^1/_2$	1	$^1/_2$	3	0.236
$^5/_2$	3	$^1/_2$	1	$^1/_2$	3	0.178
$^7/_2$	3	$^1/_2$	1	$^1/_2$	3	—0.134
$^7/_2$	4	$^1/_2$	1	$^1/_2$	3	—0.204
$^3/_2$	2	$^3/_2$	1	$^1/_2$	3	0.224
$^5/_2$	2	$^3/_2$	1	$^1/_2$	3	0.149
$^5/_2$	3	$^3/_2$	1	$^1/_2$	3	—0.141
$^7/_2$	3	$^3/_2$	1	$^1/_2$	3	—0.164
$^7/_2$	4	$^3/_2$	1	$^1/_2$	3	0.083
$^9/_2$	4	$^3/_2$	1	$^1/_2$	3	0.167
2	$^3/_2$	1	$^3/_2$	$^1/_2$	3	0.224
2	$^5/_2$	1	$^3/_2$	$^1/_2$	3	0.149
3	$^5/_2$	1	$^3/_2$	$^1/_2$	3	—0.141
3	$^7/_2$	1	$^3/_2$	$^1/_2$	3	—0.164
4	$^7/_2$	1	$^3/_2$	$^1/_2$	3	0.083
4	$^9/_2$	1	$^3/_2$	$^1/_2$	3	0.167
1	$^3/_2$	2	$^3/_2$	$^1/_2$	3	0.224
2	$^3/_2$	2	$^3/_2$	$^1/_2$	3	0.141
2	$^5/_2$	2	$^3/_2$	$^1/_2$	3	—0.141
3	$^5/_2$	2	$^3/_2$	$^1/_2$	3	—0.146

Continuation of Table 7

a	b	c	d	e	f	W
3	7/2	2	3/2	1/2	3	0.095
4	7/2	2	3/2	1/2	3	0.144
4	9/2	2	3/2	1/2	3	—0.058
5	9/2	2	3/2	1/2	3	—0.141
3/2	1	3/2	2	1/2	3	0.224
3/2	2	3/2	2	1/2	3	0.141
5/2	2	3/2	2	1/2	3	—0.141
5/2	3	3/2	2	1/2	3	—0.146
7/2	3	3/2	2	1/2	3	0.095
7/2	4	3/2	2	1/2	3	0.144
9/2	4	3/2	2	1/2	3	—0.058
1/2	1	5/2	2	1/2	3	0.236
3/2	1	5/2	2	1/2	3	0.149
3/2	2	5/2	2	1/2	3	—0.141
5/2	2	5/2	2	1/2	3	—0.141
5/2	3	5/2	2	1/2	3	0.098
7/2	3	5/2	2	1/2	3	0.134
7/2	4	5/2	2	1/2	3	—0.068
9/2	4	5/2	2	1/2	3	—0.128
1	1/2	2	5/2	1/2	3	0.236
1	3/2	2	5/2	1/2	3	0.149
2	3/2	2	5/2	1/2	3	—0.141
2	5/2	2	5/2	1/2	3	—0.141
3	5/2	2	5/2	1/2	3	0.098
3	7/2	2	5/2	1/2	3	0.134
4	7/2	2	5/2	1/2	3	—0.068
4	9/2	2	5/2	1/2	3	—0.128
5	9/2	2	5/2	1/2	3	0.043
0	1/2	3	5/2	1/2	3	0.267
1	1/2	3	5/2	1/2	3	0.178
1	3/2	3	5/2	1/2	3	—0.141
2	3/2	3	5/2	1/2	3	—0.146
2	5/2	3	5/2	1/2	3	0.098
3	5/2	3	5/2	1/2	3	0.130
3	7/2	3	5/2	1/2	3	—0.071
4	7/2	3	5/2	1/2	3	—0.121
4	9/2	3	5/2	1/2	3	0.051
5	9/2	3	5/2	1/2	3	0.114
1/2	0	5/2	3	1/2	3	0.267
1/2	1	5/2	3	1/2	3	0.178
3/2	1	5/2	3	1/2	3	—0.141

Continuation of Table 7

a	b	c	d	e	f	W
$^3/_2$	2	$^5/_2$	3	$^1/_2$	3	−0.146
$^5/_2$	2	$^5/_2$	3	$^1/_2$	3	0.098
$^5/_2$	3	$^5/_2$	3	$^1/_2$	3	0.130
$^7/_2$	3	$^5/_2$	3	$^1/_2$	3	−0.071
$^7/_2$	4	$^5/_2$	3	$^1/_2$	3	−0.121
$^9/_2$	4	$^5/_2$	3	$^1/_2$	3	0.051
$^1/_2$	0	$^7/_2$	3	$^1/_2$	3	0.267
$^1/_2$	1	$^7/_2$	3	$^1/_2$	3	−0.134
$^3/_2$	1	$^7/_2$	3	$^1/_2$	3	−0.164
$^3/_2$	2	$^7/_2$	3	$^1/_2$	3	0.095
$^5/_2$	2	$^7/_2$	3	$^1/_2$	3	0.134
$^5/_2$	3	$^7/_2$	3	$^1/_2$	3	−0.071
$^7/_2$	3	$^7/_2$	3	$^1/_2$	3	−0.119
$^7/_2$	4	$^7/_2$	3	$^1/_2$	3	0.055
$^9/_2$	4	$^7/_2$	3	$^1/_2$	3	0.109
0	$^1/_2$	3	$^7/_2$	$^1/_2$	3	0.267
1	$^1/_2$	3	$^7/_2$	$^1/_2$	3	−0.134
1	$^3/_2$	3	$^7/_2$	$^1/_2$	3	−0.164
2	$^3/_2$	3	$^7/_2$	$^1/_2$	3	0.095
2	$^5/_2$	3	$^7/_2$	$^1/_2$	3	0.134
3	$^5/_2$	3	$^7/_2$	$^1/_2$	3	−0.071
3	$^7/_2$	3	$^7/_2$	$^1/_2$	3	−0.119
4	$^7/_2$	3	$^7/_2$	$^1/_2$	3	0.055
4	$^9/_2$	3	$^7/_2$	$^1/_2$	3	0.109
5	$^9/_2$	3	$^7/_2$	$^1/_2$	3	−0.040
1	$^1/_2$	4	$^7/_2$	$^1/_2$	3	−0.204
1	$^3/_2$	4	$^7/_2$	$^1/_2$	3	0.083
2	$^3/_2$	4	$^7/_2$	$^1/_2$	3	0.144
2	$^5/_2$	4	$^7/_2$	$^1/_2$	3	−0.068
3	$^5/_2$	4	$^7/_2$	$^1/_2$	3	−0.121
3	$^7/_2$	4	$^7/_2$	$^1/_2$	3	0.055
4	$^7/_2$	4	$^7/_2$	$^1/_2$	3	0.108
4	$^9/_2$	4	$^7/_2$	$^1/_2$	3	−0.043
5	$^9/_2$	4	$^7/_2$	$^1/_2$	3	−0.100
$^1/_2$	1	$^7/_2$	4	$^1/_2$	3	−0.204
$^3/_2$	1	$^7/_2$	4	$^1/_2$	3	0.083
$^3/_2$	2	$^7/_2$	4	$^1/_2$	3	0.144
$^5/_2$	2	$^7/_2$	4	$^1/_2$	3	−0.068
$^5/_2$	3	$^7/_2$	4	$^1/_2$	3	−0.121
$^7/_2$	3	$^7/_2$	4	$^1/_2$	3	0.055
$^7/_2$	4	$^7/_2$	4	$^1/_2$	3	0.108

Continuation of Table 7

a	b	c	d	e	f	W
$^9/_2$	4	$^7/_2$	4	$^1/_2$	3	—0.043
$^3/_2$	1	$^9/_2$	4	$^1/_2$	3	0.167
$^3/_2$	2	$^9/_2$	4	$^1/_2$	3	—0.058
$^5/_2$	2	$^9/_2$	4	$^1/_2$	3	—0.128
$^5/_2$	3	$^9/_2$	4	$^1/_2$	3	0.051
$^7/_2$	3	$^9/_2$	4	$^1/_2$	3	0.109
$^7/_2$	4	$^9/_2$	4	$^1/_2$	3	—0.043
$^9/_2$	4	$^9/_2$	4	$^1/_2$	3	—0.098
1	$^3/_2$	4	$^9/_2$	$^1/_2$	3	0.167
2	$^3/_2$	4	$^9/_2$	$^1/_2$	3	—0.058
2	$^5/_2$	4	$^9/_2$	$^1/_2$	3	—0.128
3	$^5/_2$	4	$^9/_2$	$^1/_2$	3	0.051
3	$^7/_2$	4	$^9/_2$	$^1/_2$	3	0.109
4	$^7/_2$	4	$^9/_2$	$^1/_2$	3	—0.043
4	$^9/_2$	4	$^9/_2$	$^1/_2$	3	—0.098
5	$^9/_2$	4	$^9/_2$	$^1/_2$	3	0.035
2	$^3/_2$	5	$^9/_2$	$^1/_2$	3	—0.141
2	$^5/_2$	5	$^9/_2$	$^1/_2$	3	0.043
3	$^5/_2$	5	$^9/_2$	$^1/_2$	3	0.114
3	$^7/_2$	5	$^9/_2$	$^1/_2$	3	—0.040
4	$^7/_2$	5	$^9/_2$	$^1/_2$	3	—0.099
4	$^9/_2$	5	$^9/_2$	$^1/_2$	3	0.035
5	$^9/_2$	5	$^9/_2$	$^1/_2$	3	0.090
$^7/_2$	4	$^1/_2$	0	$^1/_2$	4	0.236
$^9/_2$	4	$^1/_2$	0	$^1/_2$	4	0.236
4	$^7/_2$	0	$^1/_2$	$^1/_2$	4	0.236
4	$^9/_2$	0	$^1/_2$	$^1/_2$	4	0.236
3	$^7/_2$	1	$^1/_2$	$^1/_2$	4	0.204
4	$^7/_2$	1	$^1/_2$	$^1/_2$	4	0.152
4	$^9/_2$	1	$^1/_2$	$^1/_2$	4	—0.122
5	$^9/_2$	1	$^1/_2$	$^1/_2$	4	—0.183
$^7/_2$	3	$^1/_2$	1	$^1/_2$	4	0.204
$^7/_2$	4	$^1/_2$	1	$^1/_2$	4	0.152
$^9/_2$	4	$^1/_2$	1	$^1/_2$	4	—0.122
$^5/_2$	3	$^3/_2$	1	$^1/_2$	4	0.189
$^7/_2$	3	$^3/_2$	1	$^1/_2$	4	0.122
$^7/_2$	4	$^3/_2$	1	$^1/_2$	4	—0.127
$^9/_2$	4	$^3/_2$	1	$^1/_2$	4	—0.143
3	$^5/_2$	1	$^3/_2$	$^1/_2$	4	0.189
3	$^7/_2$	1	$^3/_2$	$^1/_2$	4	0.122
4	$^7/_2$	1	$^3/_2$	$^1/_2$	4	—0.127

Continuation of Table 7

a	b	c	d	e	f	W
4	$^9/_2$	1	$^3/_2$	$^1/_2$	4	—0.143
5	$^9/_2$	1	$^3/_2$	$^1/_2$	4	0.078
2	$^5/_2$	2	$^3/_2$	$^1/_2$	4	0.183
3	$^5/_2$	2	$^3/_2$	$^1/_2$	4	0.109
3	$^7/_2$	2	$^3/_2$	$^1/_2$	4	—0.127
4	$^7/_2$	2	$^3/_2$	$^1/_2$	4	—0.124
4	$^9/_2$	2	$^3/_2$	$^1/_2$	4	0.088
5	$^9/_2$	2	$^3/_2$	$^1/_2$	4	0.128
$^5/_2$	2	$^3/_2$	2	$^1/_2$	4	0.183
$^5/_2$	3	$^3/_2$	2	$^1/_2$	4	0.109
$^7/_2$	3	$^3/_2$	2	$^1/_2$	4	—0.127
$^7/_2$	4	$^3/_2$	2	$^1/_2$	4	—0.124
$^9/_2$	4	$^3/_2$	2	$^1/_2$	4	0.088
$^3/_2$	2	$^5/_2$	2	$^1/_2$	4	0.183
$^5/_2$	2	$^5/_2$	2	$^1/_2$	4	0.105
$^5/_2$	3	$^5/_2$	2	$^1/_2$	4	—0.126
$^7/_2$	3	$^5/_2$	2	$^1/_2$	4	—0.114
$^7/_2$	4	$^5/_2$	2	$^1/_2$	4	0.091
$^9/_2$	4	$^5/_2$	2	$^1/_2$	4	0.115
2	$^3/_2$	2	$^5/_2$	$^1/_2$	4	0.183
2	$^5/_2$	2	$^5/_2$	$^1/_2$	4	0.105
3	$^5/_2$	2	$^5/_2$	$^1/_2$	4	—0.126
3	$^7/_2$	2	$^5/_2$	$^1/_2$	4	—0.114
4	$^7/_2$	2	$^5/_2$	$^1/_2$	4	0.091
4	$^9/_2$	2	$^5/_2$	$^1/_2$	4	0.116
5	$^9/_2$	2	$^5/_2$	$^1/_2$	4	—0.065
1	$^3/_2$	3	$^5/_2$	$^1/_2$	4	0.189
2	$^3/_2$	3	$^5/_2$	$^1/_2$	4	0.109
2	$^5/_2$	3	$^5/_2$	$^1/_2$	4	—0.126
3	$^5/_2$	3	$^5/_2$	$^1/_2$	4	—0.112
3	$^7/_2$	3	$^5/_2$	$^1/_2$	4	0.092
4	$^7/_2$	3	$^5/_2$	$^1/_2$	4	0.109
4	$^9/_2$	3	$^5/_2$	$^1/_2$	4	—0.069
5	$^9/_2$	3	$^5/_2$	$^1/_2$	4	—0.106
$^3/_2$	1	$^5/_2$	3	$^1/_2$	4	0.189
$^3/_2$	2	$^5/_2$	3	$^1/_2$	4	0.109
$^5/_2$	2	$^5/_2$	3	$^1/_2$	4	—0.126
$^5/_2$	3	$^5/_2$	3	$^1/_2$	4	—0.112
$^7/_2$	3	$^5/_2$	3	$^1/_2$	4	0.092
$^7/_2$	4	$^5/_2$	3	$^1/_2$	4	0.109
$^9/_2$	4	$^5/_2$	3	$^1/_2$	4	—0.069

Continuation of Table 7

a	b	c	d	e	f	W
1/2	1	7/2	3	1/2	4	0.204
3/2	1	7/2	3	1/2	4	0.122
3/2	2	7/2	3	1/2	4	—0.127
5/2	2	7/2	3	1/2	4	—0.114
5/2	3	7/2	3	1/2	4	0.092
7/2	3	7/2	3	1/2	4	0.107
7/2	4	7/2	3	1/2	4	—0.070
9/2	4	7/2	3	1/2	4	—0.102
1	1/2	3	7/2	1/2	4	0.204
1	3/2	3	7/2	1/2	4	0.122
2	3/2	3	7/2	1/2	4	—0.127
2	5/2	3	7/2	1/2	4	—0.114
3	5/2	3	7/2	1/2	4	0.092
3	7/2	3	7/2	1/2	4	0.107
4	7/2	3	7/2	1/2	4	—0.070
4	9/2	3	7/2	1/2	4	—0.102
5	9/2	3	7/2	1/2	4	0 054
0	1/2	4	7/2	1/2	4	0.236
1	1/2	4	7/2	1/2	4	0.152
1	3/2	4	7/2	1/2	4	—0.127
2	3/2	4	7/2	1/2	4	—0.124
2	5/2	4	7/2	1/2	4	0.091
3	5/2	4	7/2	1/2	4	0.109
3	7/2	4	7/2	1/2	4	—0.070
4	7/2	4	7/2	1/2	4	—0.100
4	9/2	4	7/2	1/2	4	0.056
5	9/2	4	7/2	1/2	4	0.094
1/2	0	7/2	4	1/2	4	0.236
1/2	1	7/2	4	1/2	4	0.152
3/2	1	7/2	4	1/2	4	—0.127
3/2	2	7/2	4	1/2	4	—0.124
5/2	2	7/2	4	1/2	4	0.091
5/2	3	7/2	4	1/2	4	0.109
7/2	3	7/2	4	1/2	4	—0.070
7/2	4	7/2	4	1/2	4	—0.100
9/2	4	7/2	4	1/2	4	0.056
1/2	0	9/2	4	1/2	4	0.236
1/2	1	9/2	4	1/2	4	—0.122
3/2	1	9/2	4	1/2	4	—0.143
3/2	2	9/2	4	1/2	4	0.088
5/2	2	9/2	4	1/2	4	0.116

Continuation of Table 7

a	b	c	d	e	f	W
$^5/_2$	3	$^9/_2$	4	$^1/_2$	4	—0.069
$^7/_2$	3	$^9/_2$	4	$^1/_2$	4	—0.102
$^7/_2$	4	$^9/_2$	4	$^1/_2$	4	0.056
$^9/_2$	4	$^9/_2$	4	$^1/_2$	4	0.093
0	$^1/_2$	4	$^9/_2$	$^1/_2$	4	0.236
1	$^1/_2$	4	$^9/_2$	$^1/_2$	4	—0.122
1	$^3/_2$	4	$^9/_2$	$^1/_2$	4	—0.143
2	$^3/_2$	4	$^9/_2$	$^1/_2$	4	0.088
2	$^5/_2$	4	$^9/_2$	$^1/_2$	4	0.115
3	$^5/_2$	4	$^9/_2$	$^1/_2$	4	—0.069
3	$^7/_2$	4	$^9/_2$	$^1/_2$	4	—0.102
4	$^7/_2$	4	$^9/_2$	$^1/_2$	4	0.056
4	$^9/_2$	4	$^9/_2$	$^1/_2$	4	0.093
5	$^9/_2$	4	$^9/_2$	$^1/_2$	4	—0.045
1	$^1/_2$	5	$^9/_2$	$^1/_2$	4	—0.183
1	$^3/_2$	5	$^9/_2$	$^1/_2$	4	0.078
2	$^3/_2$	5	$^9/_2$	$^1/_2$	4	0.128
2	$^5/_2$	5	$^9/_2$	$^1/_2$	4	—0.065
3	$^5/_2$	5	$^9/_2$	$^1/_2$	4	—0.106
3	$^7/_2$	5	$^9/_2$	$^1/_2$	4	0.054
4	$^7/_2$	5	$^9/_2$	$^1/_2$	4	0.094
4	$^9/_2$	5	$^9/_2$	$^1/_2$	4	—0.045
5	$^9/_2$	5	$^9/_2$	$^1/_2$	4	—0.086
5	$^9/_2$	0	$^1/_2$	$^1/_2$	5	0.213
4	$^9/_2$	1	$^1/_2$	$^1/_2$	5	0.183
5	$^9/_2$	1	$^1/_2$	$^1/_2$	5	0.135
$^9/_2$	4	$^1/_2$	1	$^1/_2$	5	0.183
$^7/_2$	4	$^3/_2$	1	$^1/_2$	5	0.167
$^9/_2$	4	$^3/_2$	1	$^1/_2$	5	0.105
4	$^7/_2$	1	$^3/_2$	$^1/_2$	5	0.167
4	$^9/_2$	1	$^3/_2$	$^1/_2$	5	0.105
5	$^9/_2$	1	$^3/_2$	$^1/_2$	5	—0.117
3	$^7/_2$	2	$^3/_2$	$^1/_2$	5	0.158
4	$^7/_2$	2	$^3/_2$	$^1/_2$	5	0.091
4	$^9/_2$	2	$^3/_2$	$^1/_2$	5	—0.115
5	$^9/_2$	2	$^3/_2$	$^1/_2$	5	—0.109
$^7/_2$	3	$^3/_2$	2	$^1/_2$	5	0.158
$^7/_2$	4	$^3/_2$	2	$^1/_2$	5	0.091
$^9/_2$	4	$^3/_2$	2	$^1/_2$	5	—0.116
$^5/_2$	3	$^5/_2$	2	$^1/_2$	5	0.154
$^7/_2$	3	$^5/_2$	2	$^1/_2$	5	0.085

Continuation of Table 7

a	b	c	d	e	f	W
$^7/_2$	4	$^5/_2$	2	$^1/_2$	5	—0.114
$^9/_2$	4	$^5/_2$	2	$^1/_2$	5	—0.098
3	$^5/_2$	2	$^5/_2$	$^1/_2$	5	0.154
3	$^7/_2$	2	$^5/_2$	$^1/_2$	5	0.085
4	$^7/_2$	2	$^5/_2$	$^1/_2$	5	—0.114
4	$^9/_2$	2	$^5/_2$	$^1/_2$	5	—0.098
5	$^9/_2$	2	$^5/_2$	$^1/_2$	5	0.085
2	$^5/_2$	3	$^5/_2$	$^1/_2$	5	0.154
3	$^5/_2$	3	$^5/_2$	$^1/_2$	5	0.083
3	$^7/_2$	3	$^5/_2$	$^1/_2$	5	—0.113
4	$^7/_2$	3	$^5/_2$	$^1/_2$	5	—0.093
4	$^9/_2$	3	$^5/_2$	$^1/_2$	5	0.086
5	$^9/_2$	3	$^5/_2$	$^1/_2$	5	0.095
$^5/_2$	2	$^5/_2$	3	$^1/_2$	5	0.154
$^5/_2$	3	$^5/_2$	3	$^1/_2$	5	0.083
$^7/_2$	3	$^5/_2$	3	$^1/_2$	5	—0.113
$^7/_2$	4	$^5/_2$	3	$^1/_2$	5	—0.093
$^9/_2$	4	$^5/_2$	3	$^1/_2$	5	0.086
$^3/_2$	2	$^7/_2$	3	$^1/_2$	5	0.158
$^5/_2$	2	$^7/_2$	3	$^1/_2$	5	0.085
$^5/_2$	3	$^7/_2$	3	$^1/_2$	5	—0.113
$^7/_2$	3	$^7/_2$	3	$^1/_2$	5	—0.091
$^7/_2$	4	$^7/_2$	3	$^1/_2$	5	0.086
$^9/_2$	4	$^7/_2$	3	$^1/_2$	5	0,091
2	$^3/_2$	3	$^7/_2$	$^1/_2$	5	0.158
2	$^5/_2$	3	$^7/_2$	$^1/_2$	5	0.085
3	$^5/_2$	3	$^7/_2$	$^1/_2$	5	—0.113
3	$^7/_2$	3	$^7/_2$	$^1/_2$	5	—0.091
4	$^7/_2$	3	$^7/_2$	$^1/_2$	5	0.086
4	$^9/_2$	3	$^7/_2$	$^1/_2$	5	0.091
5	$^9/_2$	3	$^7/_2$	$^1/_2$	5	—0.067
1	$^3/_2$	4	$^7/_2$	$^1/_2$	5	0.167
2	$^3/_2$	4	$^7/_2$	$^1/_2$	5	0.091
2	$^5/_2$	4	$^7/_2$	$^1/_2$	5	—0.114
3	$^5/_2$	4	$^7/_2$	$^1/_2$	5	—0.093
3	$^7/_2$	4	$^7/_2$	$^1/_2$	5	0.086
4	$^7/_2$	4	$^7/_2$	$^1/_2$	5	0.090
4	$^9/_2$	4	$^7/_2$	$^1/_2$	5	—0.068
5	$^9/_2$	4	$^7/_2$	$^1/_2$	5	—0.087
$^3/_2$	1	$^7/_2$	4	$^1/_2$	5	0.167
$^3/_2$	2	$^7/_2$	4	$^1/_2$	5	0.091

Continuation of Table 7

a	b	c	d	e	f	W
5/2	2	7/2	4	1/2	5	—0.114
5/2	3	7/2	4	1/2	5	—0.093
7/2	3	7/2	4	1/2	5	0.086
7/2	4	7/2	4	1/2	5	0.090
9/2	4	7/2	4	1/2	5	—0.068
1/2	1	9/2	4	1/2	5	0.183
3/2	1	9/2	4	1/2	5	0.105
3/2	2	9/2	4	1/2	5	—0.115
5/2	2	9/2	4	1/2	5	—0.098
5/2	3	9/2	4	1/2	5	0.086
7/2	3	9/2	4	1/2	5	0.091
7/2	4	9/2	4	1/2	5	—0.068
9/2	4	9/2	4	1/2	5	—0.086
1	1/2	4	9/2	1/2	5	0.183
1	3/2	4	9/2	1/2	5	0.105
2	3/2	4	9/2	1/2	5	—0.116
2	5/2	4	9/2	1/2	5	—0.098
3	5/2	4	9/2	1/2	5	0.086
3	7/2	4	9/2	1/2	5	0.091
4	7/2	4	9/2	1/2	5	—0.068
4	9/2	4	9/2	1/2	5	—0.086
5	9/2	4	9/2	1/2	5	0.055
0	1/2	5	9/2	1/2	5	0.213
1	1/2	5	9/2	1/2	5	0.135
1	3/2	5	9/2	1/2	5	—0.117
2	3/2	5	9/2	1/2	5	—0.109
2	5/2	5	9/2	1/2	5	0.085
3	5/2	5	9/2	1/2	5	0.095
3	7/2	5	9/2	1/2	5	—0.067
4	7/2	5	9/2	1/2	5	—0.087
4	9/2	5	9/2	1/2	5	0.055
5	9/2	5	9/2	1/2	5	0.081

TABLE 8

Numerical values of Racah coefficients W (*abcd*; 1*f*) for $f = 0,1,2$

a	b	c	d	e	f	W
1	0	1	0	1	0	0.577
$1/2$	$1/2$	$1/2$	$1/2$	1	0	0.500
$3/2$	$1/2$	$3/2$	$1/2$	1	0	—0.353
0	1	0	1	1	0	0.577
1	1	1	1	1	0	—0.333
2	1	2	1	1	0	0.258
$1/2$	$3/2$	$1/2$	$3/2$	1	0	—0.353
$3/2$	$3/2$	$3/2$	$3/2$	1	0	0.250
$5/2$	$3/2$	$5/2$	$3/2$	1	0	—0.204
1	2	1	2	1	0	0.258
2	2	2	2	1	0	—0.200
3	2	3	2	1	0	0.169
$3/2$	$5/2$	$3/2$	$5/2$	1	0	—0.204
$5/2$	$5/2$	$5/2$	$5/2$	1	0	0.166
$7/2$	$5/2$	$7/2$	$5/2$	1	0	—0.144
2	3	2	3	1	0	0.169
3	3	3	3	1	0	—0.142
4	3	4	3	1	0	0.125
$5/2$	$7/2$	$5/2$	$7/2$	1	0	—0.144
$7/2$	$7/2$	$7/2$	$7/2$	1	0	0.124
$9/2$	$7/2$	$9/2$	$7/2$	1	0	—0.111
3	4	3	4	1	0	0.125
4	4	4	4	1	0	—0.111
5	4	5	4	1	0	0.100
$7/2$	$9/2$	$7/2$	$9/2$	1	0	—0.111
$9/2$	$9/2$	$9/2$	$9/2$	1	0	0.100
$11/2$	$9/2$	$11/2$	$9/2$	1	0	—0.091
0	1	1	0	1	1	0.333
1	1	1	0	1	1	0.333
2	1	1	0	1	1	0.333
$1/2$	$1/2$	$1/2$	$1/2$	1	1	0.166
$3/2$	$1/2$	$1/2$	$1/2$	1	1	0.333
$1/2$	$3/2$	$1/2$	$1/2$	1	1	0.333
$3/2$	$3/2$	$1/2$	$1/2$	1	1	0.263
$1/2$	$1/2$	$3/2$	$1/2$	1	1	0.333
$3/2$	$1/2$	$3/2$	$1/2$	1	1	0.263
$1/2$	$3/2$	$3/2$	$1/2$	1	1	—0.083
$3/2$	$3/2$	$3/2$	$1/2$	1	1	—0.166
$5/2$	$3/2$	$3/2$	$1/2$	1	1	—0.250

Continuation of Table 8

a	b	c	d	e	f	W
1	0	0	1	1	1	0.333
1	1	0	1	1	1	0.333
1	2	0	1	1	1	0.333
1	0	1	1	1	1	0.333
0	1	1	1	1	1	0.333
1	1	1	1	1	1	0.166
2	1	1	1	1	1	—0.166
1	2	1	1	1	1	—0.166
2	2	1	1	1	1	—0.223
1	0	2	1	1	1	0.333
1	1	2	1	1	1	—0.166
2	1	2	1	1	1	—0.223
1	2	2	1	1	1	0.033
2	2	2	1	1	1	0.099
3	2	2	1	1	1	0.200
1/2	1/2	1/2	3/2	1	1	0.333
3/2	1/2	1/2	3/2	1	1	—0.083
1/2	3/2	1/2	3/2	1	1	0.263
3/2	3/2	1/2	3/2	1	1	—0.166
3/2	5/2	1/2	3/2	1	1	—0.250
1/2	1/2	3/2	3/2	1	1	0.263
3/2	1/2	3/2	3/2	1	1	—0.166
1/2	3/2	3/2	3/2	1	1	—0.166
3/2	3/2	3/2	3/2	1	1	—0.183
5/2	3/2	3/2	3/2	1	1	0.100
3/2	5/2	3/2	3/2	1	1	0.100
5/2	5/2	3/2	3/2	1	1	0.187
3/2	1/2	5/2	3/2	1	1	—0.250
3/2	3/2	5/2	3/2	1	1	0.100
5/2	3/2	5/2	3/2	1	1	0.187
3/2	5/2	5/2	3/2	1	1	—0.016
5/2	5/2	5/2	3/2	1	1	—0.066
7/2	5/2	5/2	3/2	1	1	—0.166
0	1	1	2	1	1	0.333
1	1	1	2	1	1	—0.166
2	1	1	2	1	1	0.033
1	2	1	2	1	1	—0.223
2	2	1	2	1	1	0.100
2	3	1	2	1	1	0.200
1	1	2	2	1	1	—0.223
2	1	2	2	1	1	0.099
1	2	2	2	1	1	0.100

Continuation of Table 8

a	*b*	*c*	*d*	*e*	*f*	*W*
2	2	2	2	1	1	0.166
3	2	2	2	1	1	—0.066
2	3	2	2	1	1	—0.066
3	3	2	2	1	1	—0.159
2	1	3	2	1	1	0.200
2	2	3	2	1	1	—0.066
3	2	3	2	1	1	—0.159
2	3	3	2	1	1	0.009
3	3	3	2	1	1	0.047
4	3	3	2	1	1	0.142
$^1/_2$	$^3/_2$	$^3/_2$	$^5/_2$	1	1	—0.250
$^3/_2$	$^3/_2$	$^3/_2$	$^5/_2$	1	1	0.100
$^5/_2$	$^3/_2$	$^3/_2$	$^5/_2$	1	1	—0.016
$^3/_2$	$^5/_2$	$^3/_2$	$^5/_2$	1	1	0.187
$^5/_2$	$^5/_2$	$^3/_2$	$^5/_2$	1	1	—0.066
$^5/_2$	$^7/_2$	$^3/_2$	$^5/_2$	1	1	—0.166
$^3/_2$	$^3/_2$	$^5/_2$	$^5/_2$	1	1	0.187
$^5/_2$	$^3/_2$	$^5/_2$	$^5/_2$	1	1	—0.066
$^3/_2$	$^5/_2$	$^5/_2$	$^5/_2$	1	1	—0.066
$^5/_2$	$^5/_2$	$^5/_2$	$^5/_2$	1	1	—0.147
$^7/_2$	$^5/_2$	$^5/_2$	$^5/_2$	1	1	0.047
$^5/_2$	$^7/_2$	$^5/_2$	$^5/_2$	1	1	0.047
$^7/_2$	$^7/_2$	$^5/_2$	$^5/_2$	1	1	0.138
$^5/_2$	$^3/_2$	$^7/_2$	$^5/_2$	1	1	—0.166
$^5/_2$	$^5/_2$	$^7/_2$	$^5/_2$	1	1	0.047
$^7/_2$	$^5/_2$	$^7/_2$	$^5/_2$	1	1	0.138
$^5/_2$	$^7/_2$	$^7/_2$	$^5/_2$	1	1	—0.005
$^7/_2$	$^7/_2$	$^7/_2$	$^5/_2$	1	1	—0.035
$^9/_2$	$^7/_2$	$^7/_2$	$^5/_2$	1	1	—0.125
1	2	2	3	1	1	0.200
2	2	2	3	1	1	—0.066
3	2	2	3	1	1	0.009
2	3	2	3	1	1	—0.159
3	3	2	3	1	1	0.047
3	4	2	3	1	1	0.142
2	2	3	3	1	1	—0.159
3	2	3	3	1	1	0.047
2	3	3	3	1	1	0.047
3	3	3	3	1	1	0.130
4	3	3	3	1	1	—0.035
3	4	3	3	1	1	—0.035
4	4	3	3	1	1	—0.121

Continuation of Table 8

a	b	c	d	e	f	W
3	2	4	3	1	1	0.142
3	3	4	3	1	1	—0.035
4	3	4	3	1	1	—0.121
3	4	4	3	1	1	0.003
4	4	4	3	1	1	0.027
5	4	4	3	1	1	0.111
$^3/_2$	$^5/_2$	$^5/_2$	$^7/_2$	1	1	—0.166
$^5/_2$	$^5/_2$	$^5/_2$	$^7/_2$	1	1	0.047
$^7/_2$	$^5/_2$	$^5/_2$	$^7/_2$	1	1	—0.005
$^5/_2$	$^7/_2$	$^5/_2$	$^7/_2$	1	1	0.138
$^7/_2$	$^7/_2$	$^5/_2$	$^7/_2$	1	1	—0.035
$^7/_2$	$^9/_2$	$^5/_2$	$^7/_2$	1	1	—0.125
$^5/_2$	$^5/_2$	$^7/_2$	$^7/_2$	1	1	0.138
$^7/_2$	$^5/_2$	$^7/_2$	$^7/_2$	1	1	—0.035
$^5/_2$	$^7/_2$	$^7/_2$	$^7/_2$	1	1	—0.035
$^7/_2$	$^7/_2$	$^7/_2$	$^7/_2$	1	1	—0.117
$^9/_2$	$^7/_2$	$^7/_2$	$^7/_2$	1	1	0.027
$^7/_2$	$^9/_2$	$^7/_2$	$^7/_2$	1	1	0.027
$^9/_2$	$^9/_2$	$^7/_2$	$^7/_2$	1	1	0.109
$^7/_2$	$^5/_2$	$^9/_2$	$^7/_2$	1	1	—0.125
$^7/_2$	$^7/_2$	$^9/_2$	$^7/_2$	1	1	0.027
$^9/_2$	$^7/_2$	$^9/_2$	$^7/_2$	1	1	0.109
$^7/_2$	$^9/_2$	$^9/_2$	$^7/_2$	1	1	—0.002
$^9/_2$	$^9/_2$	$^9/_2$	$^7/_2$	1	1	—0.022
$^{11}/_2$	$^9/_2$	$^9/_2$	$^7/_2$	1	1	—0.100
2	3	3	4	1	1	0.142
3	3	3	4	1	1	—0.035
4	3	3	4	1	1	0.003
3	4	3	4	1	1	—0.121
4	4	3	4	1	1	0.027
3	3	4	4	1	1	—0.121
4	3	4	4	1	1	0.027
3	4	4	4	1	1	0.027
4	4	4	4	1	1	0.105
5	4	4	4	1	1	—0.022
4	3	5	4	1	1	0.111
4	4	5	4	1	1	—0.022
5	4	5	4	1	1	—0.098
$^5/_2$	$^7/_2$	$^7/_2$	$^9/_2$	1	1	—0.125
$^7/_2$	$^7/_2$	$^7/_2$	$^9/_2$	1	1	0.027
$^9/_2$	$^7/_2$	$^7/_2$	$^9/_2$	1	1	—0.002
$^7/_2$	$^9/_2$	$^7/_2$	$^9/_2$	1	1	0.109

Continuation of Table 8

a	b	c	d	e	f	W
$^9/_2$	$^9/_2$	$^7/_2$	$^9/_2$	1	1	—0.022
$^7/_2$	$^7/_2$	$^9/_2$	$^9/_2$	1	1	0.109
$^9/_2$	$^7/_2$	$^9/_2$	$^9/_2$	1	1	—0.022
$^7/_2$	$^9/_2$	$^9/_2$	$^9/_2$	1	1	—0.022
$^9/_2$	$^9/_2$	$^9/_2$	$^9/_2$	1	1	—0.095
$^{11}/_2$	$^9/_2$	$^9/_2$	$^9/_2$	1	1	0.018
$^9/_2$	$^7/_2$	$^{11}/_2$	$^9/_2$	1	1	—0.100
$^9/_2$	$^9/_2$	$^{11}/_2$	$^9/_2$	1	1	0.018
$^{11}/_2$	$^9/_2$	$^{11}/_2$	$^9/_2$	1	1	0.089
1	2	1	0	1	2	0.258
2	2	1	0	1	2	0.258
3	2	1	0	1	2	0.258
$^3/_2$	$^3/_2$	$^1/_2$	$^1/_2$	1	2	0.158
$^5/_2$	$^3/_2$	$^1/_2$	$^1/_2$	1	2	0.258
$^3/_2$	$^5/_2$	$^1/_2$	$^1/_2$	1	2	0.258
$^5/_2$	$^5/_2$	$^1/_2$	$^1/_2$	1	2	0.197
$^1/_2$	$^3/_2$	$^3/_2$	$^1/_2$	1	2	0.250
$^3/_2$	$^3/_2$	$^3/_2$	$^1/_2$	1	2	0.223
$^5/_2$	$^3/_2$	$^3/_2$	$^1/_2$	1	2	0.170
$^3/_2$	$^5/_2$	$^3/_2$	$^1/_2$	1	2	—0.091
$^5/_2$	$^5/_2$	$^3/_2$	$^1/_2$	1	2	—0.149
$^7/_2$	$^5/_2$	$^3/_2$	$^1/_2$	1	2	—0.204
2	1	0	1	1	2	0.258
2	2	0	1	1	2	0.258
2	3	0	1	1	2	0.258
1	1	1	1	1	2	0.166
2	1	1	1	1	2	0.223
1	2	1	1	1	2	0.223
2	2	1	1	1	2	0.074
3	2	1	1	1	2	—0.149
2	3	1	1	1	2	—0.149
3	3	1	1	1	2	—0.178
0	1	2	1	1	2	0.258
1	1	2	1	1	2	0.223
2	1	2	1	1	2	0.152
1	2	2	1	1	2	—0.100
2	2	2	1	1	2	—0.152
3	2	2	1	1	2	—0.163
2	3	2	1	1	2	0.043
3	3	2	1	1	2	0.097
4	3	2	1	1	2	0.169
$^3/_2$	$^1/_2$	$^1/_2$	$^3/_2$	1	2	0.250

Continuation of Table 8

a	b	c	d	e	j	W
$^3/_2$	$^3/_2$	$^1/_2$	$^3/_2$	1	2	0.223
$^5/_2$	$^3/_2$	$^1/_2$	$^3/_2$	1	2	—0.091
$^3/_2$	$^5/_2$	$^1/_2$	$^3/_2$	1	2	0.170
$^5/_2$	$^5/_2$	$^1/_2$	$^3/_2$	1	2	—0.149
$^5/_2$	$^7/_2$	$^1/_2$	$^3/_2$	1	2	—0.204
$^1/_2$	$^1/_2$	$^3/_2$	$^3/_2$	1	2	0.158
$^3/_2$	$^1/_2$	$^3/_2$	$^3/_2$	1	2	0.223
$^1/_2$	$^3/_2$	$^3/_2$	$^3/_2$	1	2	0.223
$^3/_2$	$^3/_2$	$^3/_2$	$^3/_2$	1	2	0.050
$^5/_2$	$^3/_2$	$^3/_2$	$^3/_2$	1	2	—0.152
$^3/_2$	$^5/_2$	$^3/_2$	$^3/_2$	1	2	—0.152
$^5/_2$	$^5/_2$	$^3/_2$	$^3/_2$	1	2	—0.115
$^7/_2$	$^5/_2$	$^3/_2$	$^3/_2$	1	2	0.097
$^5/_2$	$^7/_2$	$^3/_2$	$^3/_2$	1	2	0.097
$^7/_2$	$^7/_2$	$^3/_2$	$^3/_2$	1	2	0.155
$^1/_2$	$^1/_2$	$^5/_2$	$^3/_2$	1	2	0.258
$^3/_2$	$^1/_2$	$^5/_2$	$^3/_2$	1	2	0.170
$^1/_2$	$^3/_2$	$^5/_2$	$^3/_2$	1	2	—0.091
$^3/_2$	$^3/_2$	$^5/_2$	$^3/_2$	1	2	—0.152
$^5/_2$	$^3/_2$	$^5/_2$	$^3/_2$	1	2	—0.152
$^3/_2$	$^5/_2$	$^5/_2$	$^3/_2$	1	2	0.050
$^5/_2$	$^5/_2$	$^5/_2$	$^3/_2$	1	2	0.106
$^7/_2$	$^5/_2$	$^5/_2$	$^3/_2$	1	2	0.146
$^5/_2$	$^7/_2$	$^5/_2$	$^3/_2$	1	2	—0.024
$^7/_2$	$^7/_2$	$^5/_2$	$^3/_2$	1	2	—0.069
$^9/_2$	$^7/_2$	$^5/_2$	$^3/_2$	1	2	—0.144
1	0	1	2	1	2	0.258
1	1	1	2	1	2	0.223
2	1	1	2	1	2	—0.100
1	2	1	2	1	2	0.152
2	2	1	2	1	2	—0.152
3	2	1	2	1	2	0.043
2	3	1	2	1	2	—0.163
3	3	1	2	1	2	0.097
3	4	1	2	1	2	0.169
1	0	2	2	1	2	0.258
0	1	2	2	1	2	0.258
1	1	2	2	1	2	0.074
2	1	2	2	1	2	—0.152
1	2	2	2	1	2	—0.152
2	2	2	2	1	2	—0.100
3	2	2	2	1	2	0.106

Continuation of Table 8

a	b	c	d	e	f	W
2	3	2	2	1	2	0,106
3	3	2	2	1	2	0,119
4	3	2	2	1	2	—0,069
3	4	2	2	1	2	—0.069
4	4	2	2	1	2	—0,136
1	0	3	2	1	2	0,258
1	1	3	2	1	2	—0,149
2	1	3	2	1	2	—0.163
1	2	3	2	1	2	0,043
2	2	3	2	1	2	0,106
3	2	3	2	1	2	0.139
2	3	3	2	1	2	—0,028
3	3	3	2	1	2	—0.078
4	3	3	2	1	2	—0.130
3	4	3	2	1	2	0.015
4	4	3	2	1	2	0,051
5	4	3	2	1	2	0,125
$^1/_2$	$^1/_2$	$^3/_2$	$^5/_2$	1	2	0,258
$^3/_2$	$^1/_2$	$^3/_2$	$^5/_2$	1	2	—0,091
$^1/_2$	$^3/_2$	$^3/_2$	$^5/_2$	1	2	0,170
$^3/_2$	$^3/_2$	$^3/_2$	$^5/_2$	1	2	—0,152
$^5/_2$	$^3/_2$	$^3/_2$	$^5/_2$	1	2	0,050
$^3/_2$	$^5/_2$	$^3/_2$	$^5/_2$	1	2	—0,152
$^5/_2$	$^5/_2$	$^3/_2$	$^5/_2$	1	2	0,106
$^7/_2$	$^5/_2$	$^3/_2$	$^5/_2$	1	2	—0,024
$^5/_2$	$^7/_2$	$^3/_2$	$^5/_2$	1	2	0.146
$^7/_2$	$^7/_2$	$^3/_2$	$^5/_2$	1	2	—0,069
$^7/_2$	$^9/_2$	$^3/_2$	$^5/_2$	1	2	—0,144
$^1/_2$	$^1/_2$	$^5/_2$	$^5/_2$	1	2	0.197
$^3/_2$	$^1/_2$	$^5/_2$	$^5/_2$	1	2	—0,149
$^1/_2$	$^3/_2$	$^5/_2$	$^5/_2$	1	2	—0.149
$^3/_2$	$^3/_2$	$^5/_2$	$^5/_2$	1	2	—0,115
$^5/_2$	$^3/_2$	$^5/_2$	$^5/_2$	1	2	0.106
$^3/_2$	$^5/_2$	$^5/_2$	$^5/_2$	1	2	0,106
$^5/_2$	$^5/_2$	$^5/_2$	$^5/_2$	1	2	0,109
$^7/_2$	$^5/_2$	$^5/_2$	$^5/_2$	1	2	—0,078
$^5/_2$	$^7/_2$	$^5/_2$	$^5/_2$	1	2	—0,078
$^7/_2$	$^7/_2$	$^5/_2$	$^5/_2$	1	2	—0,113
$^9/_2$	$^7/_2$	$^5/_2$	$^5/_2$	1	2	0,051
$^7/_2$	$^9/_2$	$^5/_2$	$^5/_2$	1	2	0,051
$^9/_2$	$^9/_2$	$^5/_2$	$^5/_2$	1	2	0,120
$^3/_2$	$^1/_2$	$^7/_2$	$^5/_2$	1	2	—0,204

Continuation of Table 8

a	b	c	d	e	f	W
$3/2$	$3/2$	$7/2$	$5/2$	1	2	0.097
$5/2$	$3/2$	$7/2$	$5/2$	1	2	0.146
$3/2$	$5/2$	$7/2$	$5/2$	1	2	—0.024
$5/2$	$5/2$	$7/2$	$5/2$	1	2	—0.078
$7/2$	$5/2$	$7/2$	$5/2$	1	2	—0.126
$5/2$	$7/2$	$7/2$	$5/2$	1	2	0.017
$7/2$	$7/2$	$7/2$	$5/2$	1	2	0.059
$9/2$	$7/2$	$7/2$	$5/2$	1	2	0.116
$7/2$	$9/2$	$7/2$	$5/2$	1	2	—0.009
$9/2$	$9/2$	$7/2$	$5/2$	1	2	—0.039
$11/2$	$9/2$	$7/2$	$5/2$	1	2	—0.111
0	1	2	3	1	2	0.258
1	1	2	3	1	2	—0.149
2	1	2	3	1	2	0.043
1	2	2	3	1	2	—0.163
2	2	2	3	1	2	0.106
3	2	2	3	1	2	—0.028
2	3	2	3	1	2	0.139
3	3	2	3	1	2	—0.078
4	3	2	3	1	2	0.015
3	4	2	3	1	2	—0.130
4	4	2	3	1	2	0.051
1	1	3	3	1	2	—0.178
2	1	3	3	1	2	0.097
1	2	3	3	1	2	0.097
2	2	3	3	1	2	0.119
3	2	3	3	1	2	—0.078
2	3	3	3	1	2	—0.078
3	3	3	3	1	2	—0.107
4	3	3	3	1	2	0.059
3	4	3	3	1	2	0.059
4	4	3	3	1	2	0.105
5	4	3	3	1	2	—0.039
2	1	4	3	1	2	0.169
2	2	4	3	1	2	—0.069
3	2	4	3	1	2	—0.130
2	3	4	3	1	2	0.015
3	3	4	3	1	2	0.059
4	3	4	3	1	2	0.113
3	4	4	3	1	2	—0.011
4	4	4	3	1	2	—0.046
5	4	4	3	1	2	—0.105

Continuation of Table 8

a	*b*	*c*	*d*	*e*	*f*	*W*
$^1/_2$	$^3/_2$	$^5/_2$	$^7/_2$	1	2	—0.204
$^3/_2$	$^3/_2$	$^5/_2$	$^7/_2$	1	2	0.097
$^5/_2$	$^3/_2$	$^5/_2$	$^7/_2$	1	2	—0.024
$^3/_2$	$^5/_2$	$^5/_2$	$^7/_2$	1	2	0.146
$^5/_2$	$^5/_2$	$^5/_2$	$^7/_2$	1	2	—0.078
$^7/_2$	$^5/_2$	$^5/_2$	$^7/_2$	1	2	0.017
$^5/_2$	$^7/_2$	$^5/_2$	$^7/_2$	1	2	—0.126
$^7/_2$	$^7/_2$	$^5/_2$	$^7/_2$	1	2	0.059
$^9/_2$	$^7/_2$	$^5/_2$	$^7/_2$	1	2	—0.009
$^7/_2$	$^9/_2$	$^5/_2$	$^7/_2$	1	2	0.116
$^9/_2$	$^9/_2$	$^5/_2$	$^7/_2$	1	2	—0.039
$^3/_2$	$^3/_2$	$^7/_2$	$^7/_2$	1	2	0.155
$^5/_2$	$^3/_2$	$^7/_2$	$^7/_2$	1	2	—0.069
$^3/_2$	$^5/_2$	$^7/_2$	$^7/_2$	1	2	—0.069
$^5/_2$	$^5/_2$	$^7/_2$	$^7/_2$	1	2	—0.113
$^7/_2$	$^5/_2$	$^7/_2$	$^7/_2$	1	2	0.059
$^5/_2$	$^7/_2$	$^7/_2$	$^7/_2$	1	2	0.059
$^7/_2$	$^7/_2$	$^7/_2$	$^7/_2$	1	2	0.101
$^9/_2$	$^7/_2$	$^7/_2$	$^7/_2$	1	2	—0.046
$^7/_2$	$^9/_2$	$^7/_2$	$^7/_2$	1	2	—0.046
$^9/_2$	$^9/_2$	$^7/_2$	$^7/_2$	1	2	—0.097
$^{11}/_2$	$^9/_2$	$^7/_2$	$^7/_2$	1	2	0.031
$^5/_2$	$^3/_2$	$^9/_2$	$^7/_2$	1	2	—0.144
$^5/_2$	$^5/_2$	$^9/_2$	$^7/_2$	1	2	0.051
$^7/_2$	$^5/_2$	$^9/_2$	$^7/_2$	1	2	0.116
$^5/_2$	$^7/_2$	$^9/_2$	$^7/_2$	1	2	—0.009
$^7/_2$	$^7/_2$	$^9/_2$	$^7/_2$	1	2	—0.046
$^9/_2$	$^7/_2$	$^9/_2$	$^7/_2$	1	2	—0.103
$^7/_2$	$^9/_2$	$^9/_2$	$^7/_2$	1	2	0.008
$^9/_2$	$^9/_2$	$^9/_2$	$^7/_2$	1	2	0.037
$^{11}/_2$	$^9/_2$	$^9/_2$	$^7/_2$	1	2	0.095
1	2	3	4	1	2	0.169
2	2	3	4	1	2	—0.069
3	2	3	4	1	2	0.015
2	3	3	4	1	2	—0.130
3	3	3	4	1	2	0.059
4	3	3	4	1	2	—0.011
3	4	3	4	1	2	0.113
4	4	3	4	1	2	—0.046
5	4	3	4	1	2	0.006
2	2	4	4	1	2	—0.136
3	2	4	4	1	2	0.051

Continuation of Table 8

a	b	c	d	e	f	W
2	3	4	4	1	2	0.051
3	3	4	4	1	2	0.105
4	3	4	4	1	2	—0.046
3	4	4	4	1	2	—0.046
4	4	4	4	1	2	—0.094
5	4	4	4	1	2	0.037
3	2	5	4	1	2	0.125
3	3	5	4	1	2	—0.039
4	3	5	4	1	2	—0.105
3	4	5	4	1	2	0.006
4	4	5	4	1	2	0.037
5	4	5	4	1	2	0.094
$^3/_2$	$^5/_2$	$^7/_2$	$^9/_2$	1	2	—0.144
$^5/_2$	$^5/_2$	$^7/_2$	$^9/_2$	1	2	0.051
$^7/_2$	$^5/_2$	$^7/_2$	$^9/_2$	1	2	—0.009
$^5/_2$	$^7/_2$	$^7/_2$	$^9/_2$	1	2	0.116
$^7/_2$	$^7/_2$	$^7/_2$	$^9/_2$	1	2	—0.046
$^9/_2$	$^7/_2$	$^7/_2$	$^9/_2$	1	2	0.008
$^7/_2$	$^9/_2$	$^7/_2$	$^9/_2$	1	2	—0.103
$^9/_2$	$^9/_2$	$^7/_2$	$^9/_2$	1	2	0.037
$^{11}/_2$	$^9/_2$	$^7/_2$	$^9/_2$	1	2	—0.005
$^5/_2$	$^5/_2$	$^9/_2$	$^9/_2$	1	2	0.120
$^7/_2$	$^5/_2$	$^9/_2$	$^9/_2$	1	2	—0.039
$^5/_2$	$^7/_2$	$^9/_2$	$^9/_2$	1	2	—0.039
$^7/_2$	$^7/_2$	$^9/_2$	$^9/_2$	1	2	—0.097
$^9/_2$	$^7/_2$	$^9/_2$	$^9/_2$	1	2	0.037
$^7/_2$	$^9/_2$	$^9/_2$	$^9/_2$	1	2	0.037
$^9/_2$	$^9/_2$	$^9/_2$	$^9/_2$	1	2	0.087
$^{11}/_2$	$^9/_2$	$^9/_2$	$^9/_2$	1	2	—0.030
$^7/_2$	$^5/_2$	$^{11}/_2$	$^9/_2$	1	2	—0.111
$^7/_2$	$^7/_2$	$^{11}/_2$	$^9/_2$	1	2	0.031
$^9/_2$	$^7/_2$	$^{11}/_2$	$^9/_2$	1	2	0.095
$^7/_2$	$^9/_2$	$^{11}/_2$	$^9/_2$	1	2	—0.005
$^9/_2$	$^9/_2$	$^{11}/_2$	$^9/_2$	1	2	—0.030
$^{11}/_2$	$^9/_2$	$^{11}/_2$	$^9/_2$	1	2	—0.086

TABLE 9

Numerical values of the coefficients $Z^2\ (l_1J_1l_2J_2;\ ½f)$ for $f = 0, 1, 2, 3, 4, 5$ The positive sign is retained before a number without an asterisk and the negative sign before a number with an asterisk, when extracting the square root.

$Z^2\ (l_1J_1l_2J_2;\ {}^1/_2,\ 0)$

J_2	l_2	J_1 = 1/2, l_1 = 0	1/2, 1	3/2, 1	3/2, 2	5/2, 2	5/2, 3	7/2, 3	7/2, 4	9/2, 4	9/2, 5
1/2	0	2									
	1		2								
3/2	1			*4							
	2				*4						
5/2	2					6					
	3						6				
7/2	3							*8			
	4								*8		
9/2	4									10	
	5										10

$Z^2\ (l_1J_1l_2J_2;\ {}^1/_2,\ 1)$

J_2	l_2	J_1 = 1/2, l_1 = 0	1/2, 1	3/2, 1	3/2, 2	5/2, 2	5/2, 3	7/2, 3	7/2, 4	9/2, 4	9/2, 5
1/2	0		2	4							
	1	*2			4						
3/2	1	*4			*4/5	*36/5					
	2		*4	4/5			*36/5				
5/2	2			36/5			18/35	72/7			
	3				36/5	*18/35			72/7		
7/2	3					*72/7			*8/21	*40/3	
	4						*72/7	8/21			*40/3
9/2	4							40/3			10/33
	5								40/3	*10/33	

Continuation of Table 9

J_2	l_2	J_1 = 1/2		3/2		5/2		7/2		9/2	
		l_1 = 0	1	1	2	2	3	3	4	4	5
		$Z^2(l_1J_1l_2J_2;\ 1/2,\ 2)$									
1/2	0				4	6					
	1			*4			6				
3/2	1		*4	*4			*12/7	*72/7			
	2	4			*4	12/7			*72/7		
5/2	2	6			12/7	48/7			8/7	100/7	
	3		6	*12/7			48/7	*8/7			100/7
7/2	3			*72/7			*8/7	*200/21			*200/231
	4				*72/7	8/7			*200/21	200/231	
9/2	4					100/7			200/231	400/33	
	5						100/7	*200/231			400/33

$Z^2(l_1J_1l_2J_2;\ 1/2,\ 3)$

J_2	l_2	J_1 = 1/2, l_1 = 0	1/2, 1	3/2, 1	3/2, 2	5/2, 2	5/2, 3	7/2, 3	7/2, 4	9/2, 4	9/2, 5
1/2	0						6	8			
	1					*6			8		
3/2	1				*36/5	*24/5			*8/3	*40/3	
	2			36/5			*24/5	8/3			*40/3
5/2	2		6	24/5			16/5	8			20/11
	3	*6			24/5	*16/5			8	*20/11	
7/2	3	*8			*8/3	*8			*24/11	*120/11	
	4		*8	8/3			*8	24/11			*120/11
9/2	4			40/3			20/11	120/11			240/143
	5				40/3	*20/11			120/11	*240/143	

Continuation of Table 9

J_2	l_2	J_1 = 1/2		3/2		5/2		7/2		9/2	
		l_1 = 0	1	1	2	2	3	3	4	4	5
		$Z^2(l_1J_1l_2J_2$; 1/2, 4)									
1/2	0								8	10	
	1							*8			10
3/2	1						*72/7	*40/7			*40/11
	2					72/7			*40/7	40/11	
5/2	2				72/7	36/7			360/77	720/77	
	3			*72/7			36/7	*360/77			720/77
7/2	3		*8	*40/7			*360/77	*648/77			*3240/1001
	4	8			*40/7	360/77			*648/77	3240/1001	
9/2	4	10			40/11	720/77			3240/1001	1620/143	
	5		10	*40/11			720/77	*3240/1001			1620/143
		$Z^2(l_1J_1l_2J_2$; 1/2, 5)									
1/2	0										10
	1									*10	
3/2	1								*40/3	*20/3	
	2							40/3			*20/3
5/2	2						100/7	40/7			80/13
	3					*100/7			40/7	*80/13	
7/2	3				*40/3	*40/7			*600/91	*120/13	
	4			40/3			*40/7	600/91			*120/13
9/2	4		10	20/3			80/13	120/13			60/13
	5	*10			20/3	*80/13			120/13	*60/13	

TABLE 10

Numerical values of the coefficients $Z^2\ (l_1 J_1 l_2 J_2;\ 1f)$ for $f = 0, 1, 2, 3, 4, 5$

The positive sign is retained before a number without an asterisk and the negative sign before a number with an asterisk, when extracting the square root.

$Z^2\ (l_1\ J_1\ l_2\ J_2;\ 1,\ 0)$

J_2	l_2	J_1 = 0, l_1 = 1	J_1 = 1, l_1 = 0	J_1 = 1, l_1 = 1	J_1 = 1, l_1 = 2	J_1 = 2, l_1 = 1	J_1 = 2, l_1 = 2	J_1 = 2, l_1 = 3	J_1 = 3, l_1 = 2	J_1 = 3, l_1 = 3	J_1 = 3, l_1 = 4	J_1 = 4, l_1 = 3	J_1 = 4, l_1 = 4	J_1 = 4, l_1 = 5
0	1	*1												
	0		3											
1	1			3										
	2				3									
	1					*5								
2	2						*5							
	3							*5						
	2								7					
3	3									7				
	4										7			
	3											*9		
4	4												*9	
	5													*9

$Z^2\ (l_1\ J_1\ l_2\ J_2;\ 1,\ 1)$

J_2	l_2	J_1 = 0, l_1 = 1	J_1 = 1, l_1 = 0	J_1 = 1, l_1 = 1	J_1 = 1, l_1 = 2	J_1 = 2, l_1 = 1	J_1 = 2, l_1 = 2	J_1 = 2, l_1 = 3	J_1 = 3, l_1 = 2	J_1 = 3, l_1 = 3	J_1 = 3, l_1 = 4	J_1 = 4, l_1 = 3	J_1 = 4, l_1 = 4	J_1 = 4, l_1 = 5
0	1		*1		*2									
	0	1		3		5								
1	1		3*		3/2		9/2							
	2	2		*3/2		1/10		27/5						
	1		*5		*1/10		*3/2		*42/5					
2	2			*9/2		3/2		*1		8*				
	3				*27/5		1		*1/35		*60/7			
	2					42/5		1/35		1		81/7		
3	3						8		*1		3/4		45/4	
	4							60/7		*3/4		1/84		35/3
	3								*81/7		*1/84		*3/4	
4	4									*45/4		3/4		*3/5
	5										*35/3		3/5	

Continuation of Table 10

$Z^2\ (l_1\ J_1\ l_2\ J_2;\ 1,\ 2)$

J_2	l_2	J_1 = 0	1			2			3			4		
	l_1	1	0	1	2	1	2	3	2	3	4	3	4	5
0	1					*2		*3						
1	0				3		5		7					
	1			*3/2		*9/2		3		6				
	2		3		3/2		*5/2		2/7		54/7			
2	1	*2		*9/2		*7/2		*3/7		*3		*81/7		
	2		5		*5/2		*25/14		20/7		*15/7		*75/7	
	3	*3		3		*3/7		*32/7		2		*2/21		*250/2
3	2		7		2/7		20/7		48/7		1/7		15/7	
	3			6		*3		2		21/4		*25/12		5/3
	4				54/7		*15/7		1/7		625/84		*45/28	
4	3					*81/7		*2/21		*25/12		*275/28		*5/77
	4						*75/7		15/7		*45/28		*2601/308	
	5							*250/21		5/3		*5/77		*112/11

$Z^2\ (l_1\ J_1\ l_2\ J_2;\ 1,\ 3)$

J_2	l_2	J_1 = 0	1			2			3			4		
	l_1	1	0	1	2	1	2	3	2	3	4	3	4	5
0	1								*3		*4			
1	0							5		7		9		
	1						*3		*6		9/2		15/2	
	2					27/5		8/5		*7/2		1/2		10
2	1				*27/5		*6		*18/5		*5/6		*9/2	
	2			3		6		*4		*7/6		25/6		*10/3
	3		*5		*8/5		4		*16/15		*5		3	
3	2	3		6		18/5		16/15		14/3		22/3		10/33
	3		*7		7/2		7/6		*14/3		7/2		105/22	
	4	4		*9/2		5/6		5		*7/2		9/22		90/11
4	3		*9		*1/2		*25/6		*22/3		*9/22		*81/22	
	4			*15/2		9/2		*3		*105/22		81/22		*162/55
	5				*10		10/3		*10/33		*90/11		162/55	

Continuation of Table 10

$Z_2\ (l_1\ J_1\ l_2\ J_2;\ 1,\ 4)$

J_2	l_2	J_1 = 0	1			2			3			4		
		l_1 = 1	0	1	2	1	2	3	2	3	4	3	4	5
0	1											*4		*5
1	0										7		9	
	1									*9/2		*15/2		6
	2								54/7		25/14		*9/2	
2	1							*60/7		*15/2		*55/14		*14/11
	2						40/7		50/7		*75/14		*135/154	
	3					*60/7		*10/7		5		*135/77		*432/77
3	2				54/7		50/7		22/7		150/77		486/77	
	3			*9/2		*15/2		5		7/22		*135/22		54/11
	4		7		25/14		*75/14		150/77		729/154		*729/154	
4	3	*4		*15/2		*55/14		*135/77		*135/22		*1053/154		*810/1001
	4		9		*9/2		*135/154		486/77		*729/154		*6561/2002	
	5	*5		6		*14/11		*432/77		54/11		*810/1001		*1134/143

$Z^2\ (l_1\ J_1\ l_2\ J_2;\ 1,\ 5)$

J_2	l_2	J_1 = 0	1			2			3			4		
		l_1 = 1	0	1	2	1	2	3	2	3	4	3	4	5
0	1													
1	0													9
	1												*6	
	2											10		2
2	1										*35/3		*9	
	2									25/3		25/3		*20/3
	3								*250/21		*10/7		6	
3	2							250/21		25/3		65/21		112/39
	3						*25/3		*25/3		25/4		3/52	
	4					35/3		10/7		*25/4		1125/364		63/13
4	3				*10		*25/3		*65/21		*1125/364		*405/52	
	4			6		9		*6		*3/52		405/52		*81/13
	5		*9		*2		20/3		*112/39		*63/13		81/13	

RESOLUTION OF A PLANE WAVE

1. The wave function of the photon (23) can be written in the following compact form[1]:

$$\vec{A}_{Lp}^{m_L} = -\sqrt{2}\sum_l C_{L-1,\,11}^{l0}\,\delta(l,\ p)\, i^{l-L-p}\,\vec{\Phi}_{Ll}^{m_L}, \qquad \text{(A.13)}$$

where (l,p) differs from zero for $l = L \pm 1$ (when $p = 1$) and for $l = L$ (when $p = 0$).

2. To resolve a plane wave into the functions (2.4), it is necessary first of all, to use the expansion of the plane wave of a spin-less particle in terms of spherical Bessel functions (12):

$$\exp(ikz) = \sqrt{4\pi}\sum_{l=0}^{\infty} i^l\,(2l+1)^{1/2}\left(\frac{\pi}{2kr}\right)^{1/2} J_{l+1/2}(kr)\,Y_{l0}(\vartheta) =$$

$$\underset{r\to\infty}{=} \frac{i\sqrt{\pi}}{kr}\sum_{l=0}^{\infty} i^l\,(2l+1)^{1/2}Y_{l0}\left\{\exp\left[-i\left(kr-\frac{\pi l}{2}\right)\right] - \exp\left[i\left(kr-\frac{\pi l}{2}\right)\right]\right\}. \qquad \text{(A.13a)}$$

When the spins of the photon and nucleon are taken into consideration, the incident wave in the entry channel α will be

---

1 Here it should be borne in mind that

$$C_{L-1,\,11}^{L0} = -\frac{1}{\sqrt{2}}, \quad C_{L-1,\,11}^{L+1,\,0} = \sqrt{\frac{L}{2(2L+1)}}, \quad C_{L-1,\,11}^{L-1,\,0} = \sqrt{\frac{L+1}{2(2L+1)}}.$$

defined by the expression

$$\exp(ikz)\vec{\chi}_m \chi_I^{m_I}, \tag{A.13b}$$

Here $m = \pm 1$, since the photon moves along the z-axis and only two transverse polarizations of the photons are possible.

Hence,

$$\exp(ikz)\vec{\chi}_m = \frac{i\sqrt{\pi}}{kr}\sum_{l=0}^{\infty} i^l (2l+1)^{1/2} \sum_{L=l-1}^{l+1} C_{l01m}^{Lm_L} \left[\vec{\Phi}_{Ll}^{m_L(-)} - \vec{\Phi}_{Ll}^{m_L(+)}\right]. \tag{A.14}$$

Since

$$\sqrt{2}\,(2l+1)\,C_{l01m}^{Lm_L} = \begin{cases} \sqrt{2L+1}\sqrt{\dfrac{L+1}{2L+1}} & \text{for } L = l+1; \\ \sqrt{2L+1}\sqrt{\dfrac{L}{2L+1}} & \text{for } L = l-1; \\ -\sqrt{2L+1}\cdot m_L & \text{for } L = l; \\ 0 & \text{for the remaining } L, \end{cases} \tag{A.14a}$$

the function $\vec{\Phi}_{Ll}^{m_L}$ in the expression (A.14) can be replaced by $\vec{A}_{Lp}^{m_L}$: according to (A.13):

$$\exp(ikz)\vec{\chi}_m = \frac{i}{kr}\sqrt{\frac{\pi}{2}}\sum_{Lp} i^{L+p}(2L+1)^{1/2} m^p \left[\vec{A}_{Lp}^{m_L(-)} - \vec{A}_{Lp}^{m_L(+)}\right]. \tag{A.15}$$

Then, taking (2.4) into account, we finally get

$$\exp(ikz)\vec{\chi}_m \chi_{m_I} = \frac{i}{kr}\sqrt{\frac{\pi}{2}}\sum_{JLp} i^{L+p}(2L+1)^{1/2} m^p C_{Lm_L,\,Im_I}^{JM} \times \left[\vec{\Phi}_{JLp}^{\prime M(-)} - \vec{\Phi}_{JLp}^{\prime M(+)}\right]. \tag{A.16}$$

APPENDIX 4

DERIVATION OF THE EXPRESSION FOR THE POLARIZATION OF RECOIL NUCLEONS (20)

We shall find an expression for

$$P \doteq \frac{d\sigma_+ - d\sigma_-}{d\sigma_+ + d\sigma_-}. \qquad \text{(A. 16a)}$$

To calculate $d\sigma_+$ and $d\sigma_-$ we make use of the same method as was used to obtain the expression (2.18) for the differential cross-section of photo-production of mesons. The only difference is that summation is not carried out over the final projections of the spin of the nucleon, the cross-sections for the two projections of the spin being calculated separately, which gives $d\sigma_+$ and $d\sigma_-$.

We shall start from the expression (2.13). In this expression $m_{s'} = \frac{1}{2}$ corresponds to the direction of the spin of the recoil nucleon along the z-axis (let this state be described by the function α) and $m_{s'} = -\frac{1}{2}$ – opposite to the z-axis (this state is described by β).

Let us assume for simplicity of calculation that the incident photon moves along the z-axis and the nucleon formed flies out into the xz-plane. Then the polarization of nucleons along the y-axis will be different from zero. The functions α and β are related in the following way with the functions γ and δ, which characterize the state of the nucleon with spins directed along the y-axis and opposite to it:

$$f_{\pm} = \sum i^{L-l'+p} \frac{\sqrt{\pi}}{2} (2L+1)^{1/2} C^{JM}_{Lm,\, Im_I} S Y_{l'\mu'}(\vartheta) \times \times \left[C^{JM}_{l'\mu',\, 1/2 1/2} \mp C^{JM}_{l'\mu',\, 1/2 -1/2}\right]. \quad \text{(A. 16b)}$$

If these relationships are taken into account, the expressions for the amplitudes f_+ and f_- of the photo-production of mesons, accompanied by the formation of recoil nucleons with spins directed along the y-axis and against it, can be written as:

$$\alpha = \frac{1}{\sqrt{2}}(\gamma + \delta), \qquad \beta = -\frac{i}{\sqrt{2}}(\gamma - \delta). \quad \text{(A. 16c)}$$

Hence

$$d\sigma_{\pm} = \frac{1}{4k^2} \sum_{mm_I} |f_{\pm}|^2, \quad \text{(A. 17)}$$

where k is the wave number of the photon.

In the expressions for $d\sigma_+$ and $d\sigma_-$ there are terms containing products of Clebsch-Gordan coefficients with identical signs of the projections m_s, (we shall denote them by a_+ and a_-), as well as terms with different signs of m_s, (i.e. b_+ and b_-). In this case $a_+ = a_- \equiv a$ and $b_+ = - b_-$, i.e.

$$d\sigma_+ = a\, d\Omega + d\sigma'; \qquad d\sigma_- = a\, d\Omega - d\sigma', \quad \text{(A. 17')}$$

where

$$d\sigma' = b_+ d\Omega. \quad \text{(A. 17' a)}$$

Using relation (A.6) we find

$$d\sigma' = \frac{1}{16k^2} \sum_{\substack{J_1L_1p_1l_1' \\ J_2L_2p_2l_2'n}} i^{L_1-L_2-l_1'+l_2'+p_1-p_2} \sqrt{\pi} \times$$

$$\times \left[(2L_1+1)(2L_2+1)(2l_1'+1)(2l_2'+1)\frac{1}{2n+1}\right]^{1/2} \times$$

$$\times \operatorname{Im}\{S^*(1)\,S(2)\} \sum_{mm_I} (-1)^{\mu_2'} m^{p_1+p_2} C^{J_1M_1}_{L_1m,\,Im_I} C^{J_2M_2}_{L_2m,\,Im_I} \times \tag{A.18}$$

$$\times C^{n0}_{l_1'0,\,l_2'0} C^{n\mu_1'-\mu_2'}_{l_1'\mu_1',\,l_2'-\mu_2'} \left(C^{J_1M_1}_{l_1'\mu_1',\,{}^1/_2{}^1/_2} C^{J_2M_2}_{l_2'\mu_2',\,{}^1/_2-{}^1/_2} -\right.$$

$$\left. - C^{J_1M_1}_{l_1'\mu_1',\,{}^1/_2-{}^1/_2} C^{J_2M_2}_{l_2'\mu_2',\,{}^1/_2{}^1/_2}\right) Y_n^{\mu_1'-\mu_2'}(\vartheta)\, d\Omega.$$

Substituting (A.17'). in the expression for P and bearing in mind that $d\sigma_+ + d\sigma_- = d\sigma$, we shall get

$$P = \frac{2d\sigma'}{d\sigma}. \tag{A.19}$$

The quantity P can be determined with the help of (A.19) but this leads to rather cumbrous calculations. The calculations are considerably simplified if the summation is carried out over the projections of the spins m and m_I in the expression (A.19).

To carry out the above summation we use the following relationships (see Appendix 2):

$$C^{e\alpha+\beta}_{a\alpha,\,b\beta} C^{c\alpha+\beta+\delta}_{e\alpha+\beta,\,d\delta} = \sum_f \sqrt{(2e+1)(2f+1)}\, C^{c\alpha+\beta+\delta}_{a\alpha,\,f\beta+\delta} C^{f\beta+\delta}_{b\beta,\,d\delta} W(abcd;\,ef)$$

and

$$\sum_\beta C^{e\alpha+\beta}_{a\alpha,\,b\beta} C^{c\gamma}_{c\alpha+\beta,\,d\gamma-\alpha-\beta} C^{f\gamma-\alpha}_{b\beta,\,d\gamma-\alpha-\beta} = \sqrt{(2e+1)(2f+1)}\, C^{c\gamma}_{a\alpha,\,f\gamma-\alpha}(Wabcd;\,ef). \tag{A.19a}$$

Using the symmetry properties of the Clebsch-Gordan coefficients and the relationships just given, we find:

$$C^{J_1M_1}_{l_1'\mu_1', {}^1/_2{}^1/_2}\, C^{n\mu_1'-\mu_2'}_{l_1'\mu_1',\, l_2'-\mu_2'} =$$

$$= -\sqrt{2J_1+1}\sum_f \sqrt{2f+1}\, C^{fM_1-\mu_2'}_{J_1M_1,\, l_2'-\mu_2'}\, C^{n\mu_1'-\mu_2'}_{{}^1/_2-{}^1/_2,\, fM_1-\mu_2'} W\left({}^1/_2 J_1 n l_2';\, l_1' f\right);$$

$$C^{fM_1-\mu_2'}_{J_1M_1,\, l_2'-\mu_2'}\, C^{J_2M_2}_{l_2'\mu_2',\, {}^1/_2-{}^1/_2} = (-1)^{f-\mu_2'-M_1+1}\sqrt{(2f+1)(2J_2+1)\frac{1}{2}}\times$$

$$\times\sum_z \sqrt{2z+1}\, C^{zM_1-M_2}_{J_1M_1,\, J_2-M_2}\, C^{{}^1/_2{}^1/_2}_{f\mu_2'-M_1,\, zM_1-M_2} W\left(fJ_1{}^1/_2 J_2;\, l_2' z\right);$$

$$\sum_{m_I} C^{J_1M_1}_{L_1m,\, Im_I}\, C^{J_2M_2}_{L_2m,\, Im_I}\, C^{zM_1-M_2}_{J_1M_1,\, J_2-M_2} =$$

$$= (-1)^{L_2-m-M_2+J_2}\sqrt{(2J_1+1)(2J_2+1)}\cdot C^{z0}_{L_1mL_2-m} W\left(L_1 I z J_2;\, J_1 L_2\right).$$

Substituting these expressions in (A.18) and summing over m and f, we get

$$d\sigma' = \frac{1}{4k^2}\sum i^{L_1-l_1'+p_1-L_2+l_2'-p_2}\times$$

$$\times\sqrt{\frac{\pi}{2}(2L_1+1)(2L_2+1)(2l_1'+1)(2l_2'+1)}\times$$

$$\times(2J_1+1)(2J_2+1)\sqrt{\frac{2n(n+1)}{2n+1}}\, C^{n0}_{l_1'0l_2'0}\, C^{n0}_{L_11,\, L_2-1} W\left(L_1J_1L_2J_2;\, {}^1/_2 n\right)\times$$

$$\times\left[W\left({}^1/_2 l_1' n - {}^1/_2 l_2';\, J_1 n\right) W\left(l_2' J_1{}^1/_2 n;\, n-{}^1/_2 J_2\right) -\right.$$

$$\left.- W\left({}^1/_2 l_1' n + {}^1/_2 l_2' J_1 n\right) W\left(l_2' J_1{}^1/_2 n;\, n+{}^1/_2 J_2\right)\right] \operatorname{Im}\left[S^*(1)\, S(2)\right] Y_{n-1} d\Omega. \tag{A.20}$$

In this expression the summation is carried out over J_1, J_2, L_1, L_2, p_1, p_2, l'_1, l'_2 and n.

Here it is considered that $z = n$ and that the presence of two components in the round brackets of the expression (A.18) leads to doubling of the final result.

For the case when mesons are formed only in s- and p-states, we get expression (4.13), with the help of (A.20) and taking (A.19) into consideration, for the polarization of recoil nucleons.

(The substitution mentioned in the footnote on p. 19 must also be introduced in the expression obtained.)

REFERENCES

1. H. Bethe and F. Hoffman. Mesons and fields. vol. 2. 1957.
2. L.M. Barkov and B.A. Nikol'skii. π -mesons. Usp. Fiz. Nauk, 61, 341, 1957.
3. K. Watson, J. Keck, A. Tollestrup and R. Walker. Partial wave analysis of the experimental photomeson cross sections. Phys. Rev., 101, 1159, 1956.
4. V.P. Silin and V.Ya. Fainberg. Scattering of π-mesons on nucleons, Usp. Fiz. Nauk. 50, 325, 1953.
5. L. Van Hove. Coulomb effects in pion scattering by protons under the charge independence hypothesis. Phys. Rev., 88, 1358, 1952.
6. Problems of present-day physics. No. 3, Reports of the Seventh Rochester Conference.
7. H.L. Anderson, W.C. Davidon and U.E. Kruse. Causality in the pion-proton scattering. Phys. Rev., 100, 339, 1955.
   A.M. MUKHIN and B.M. Pontecorvo. Scattering of π-mesons on nucleons, Zhur. Exp. Teor. Fiz., 31, 550, 1956; Problems of present-day physics, No. 2, 1957.
8. D. Oakley and R. Walker. Photoproduction of neutral pions in hydrogen; magnetic analysis of recoil protons. Phys. Rev., 97, 1283, 1955.
9. J. Blatt and L. Biedenharn. The angular distribution of scattering and reaction cross sections. Rev. Mod. Phys., 24, 258, 1952.
10. M. Morita, A. Sugie and S. Yoshida. Note on the angular distribution of photo-reaction. Progr. Theor. Phys., 12, 713, 1954.
11. D.I. Blokhintsev. Fundamentals of Quantum Mechanics. (Osnovy kvantovoi mekhaniki). Gostekhizdat, 1949.
12. L.D. Landau and E.M. Lifshitz. Quantum Mechanics (Kvantovaya Mekhanika). Part 1, Gostekhizdat, 1948. (English translation published by Pergamon Press. 1959).

13. K.A. Ter-Martiros'yan. Zhur. Exp. Teor. Fiz., 21, 895, 1951.
14. B. Feld. The angular and energy distributions in photomeson production. Phys. Rev., 89, 330, 1953; K. Watson. Some general relations between the photoproduction and scattering of π-mesons. Phys. Rev., 95, 228, 1954.
15. N.F. Nelipa. Zhur. Exp. Teor. Fiz. 33, 271, 1957.
16. K. Watson. The hypothesis of charge independence for nuclear phenomena. Phys. Rev. 85, 852, 1952; A.M. Baldin and V.V. Mikhailov. Dok. Akad. Nauk. SSSR, 91, 479, 1953.
17. D. Blatt and V. Weisskopf. Theoretical nuclear physics.
18. M. Gell-Mann and K.M. Watson. Usp. Fiz. Nauk, 59, 399, 1956.
19. E. Fermi. Lectures on π-mesons and nucleons.
20. N.F. Nelipa and V.A. Feoktistov. Zhur. Exp. Teor. Fiz., 35, 244, 1958.
21. V.I. Ritus. Theory of reactions with polarized particles and γ-quanta. Thesis. Physical Institute, Acad. Sci., USSR, 1958.
22. G.Ya. Lyubarskii. Theory of groups and its application to physics (Teoriya grupp i yeyo primeneniye v fizike). Gostekhizdat, Moscow, 1957.; L. Bidenharn. Some properties of the Racah- and associated-coefficients, Rev. Mod. Phys. 24, 249, 1952.
23. A.I. Akhiezer and V.B. Berestetskii. Quantum electrodynamics (Kvantovaya elektrodinamika) Gostekhizdat, 1953.
24. A. Simon and J. Van der Sluis. Table of Racah-coefficients. Oak Ridge National Laboratory ORNL-1679, 1954.
25. L. Biedenharn. Table of Z-coefficients. Oak Ridge National Laboratory, ORNL-1501, 1953.
26. E. Condon and G. Shortley. Theory of atomic spectra.
27. R. Saito and M. Morita. Clebsch-Gordan Coefficients for $j_2 = 5/2$. Progr. Theor. Phys. 13, 540, 1955.
28. D.L. Falkoff, G.S. Colladay and R.E. Sells. Transformation amplitude for vector addition of angular momentum. Can. Journ. Phys. 30, 253, 1952.
29. M. Yamada and M. Morita. On the β-ray angular correlations. Progr. Theor. Phys., 8, 431, 1952.
30. A. Simon. Table of Clebsch-Gordan Coefficients. Oak Ridge National Laboratory, ORNL-1718, 1954.